Sketching Splendor

AMERICAN NATURAL HISTORY, 1750–1850

Anna Majeski

With an Introduction by
Michelle Craig McDonald

Edited by the

AMERICAN
PHILOSOPHICAL
SOCIETY
Library
&Museum

Published on the occasion of the exhibition
Sketching Splendor: American Natural History, 1750–1850
April 12–December 29, 2024

American Philosophical Society
104 South Fifth Street
Philadelphia, PA 19106
amphilsoc.org

EDITED BY the American Philosophical Society

PROJECT MANAGEMENT Mary Grace Wahl

DESIGN barb barnett graphic design llc

PRINTING Brilliant Graphics, Exton, PA

Front cover: *Sandhill Cranes* (detail), Titian Ramsay Peale, 1820, watercolor and graphite on paper. APS.

Identifiers: ISBN 9781606180402 | Library of Congress Control Number: 2024936980

Also available as a free downloadable PDF at:
https://www.amphilsoc.org/museum/exhibitions

Contents

Introduction

Sketching Splendor: American Natural History, 1750–1850 brings together ideas and images from three of North America's best-known early naturalists, William Bartram (1739–1823), Titian Ramsey Peale (1799–1885), and John James Audubon (1785–1851). Their careers spanned a period of extraordinary growth in the study of the natural sciences as well as a momentous time in a new nation's history. Even before U.S. independence, North American colonists had pushed to expand land available for settlement. Once free from Britain's imperial strictures, both private investors and the new federal government focused increasingly on the West and South, determined to study and understand, but in particular control, the continent and its resources.

The goals of these early scientists cannot be disentangled from the political agendas that often prompted and financially supported their expeditions. Vast interior landscapes beckoned, and ventures into them promised new routes to wealth that mirrored, in many ways, the first European forays across the Atlantic Ocean centuries before. But instead of boarding ships, these new voyagers traveled overland, bearing scrolls of parchment, paints and brushes, and myriad other tools that documented what they encountered. For to know what was possible in North America, it was first important to know what was there.

The scientists whose studies frame this exhibition—Bartram, Peale, and Audubon—provided a wealth of scientifically detailed and visually compelling information from parts of the continent rarely, if ever, visited by non-indigenous people. In their sketches and paintings, they documented the animals and plants they encountered, as well as their ecological habitats, while describing the nations and communities that lived there.

Philadelphia's scientific community generally, and the American Philosophical Society (APS) in particular, supported the work of these naturalists. Bartram was the son of John Bartram, who helped found the APS in 1743. Peale's father, Charles Willson Peale, displayed his specimens in a museum housed in our Philosophical Hall, while Audubon, who aspired to describe and paint every bird of America, became an APS member in 1831, and Peale in 1833.

The histories of these men have, of course, already been widely studied. This exhibition tries to move beyond their individual exploits to tell the stories of others—female and male, enslaved and free, Native American, African, and European—involved in these expeditions. Indeed, this tension between undeniable accomplishment and historical erasure lies at the heart of *Sketching Splendor*. Visitors will meet Maria Martin, the sister-in-law of John Bachman, one of Audubon's most frequent collaborators and a self-taught artist, whose studies of birds and natural backgrounds were so sophisticated that Audubon included 30 images in his publications (although he only credited her for nine!). Museumgoers will also be introduced to Thomas, an enslaved man in Bachman's household whose interest in the natural world had apparently been stimulated by Audubon's repeated stays at his enslaver's home. Indeed, Thomas's responsibilities for Bachman's aviary and his own experiments with taxidermy may well have provided the models for some of Audubon's paintings. Unfortunately, not all contributors to the work of Bartram, Peale, and Audubon can be identified or acknowledged, but the science with which the three men are credited was clearly a collaborative endeavor.

The naturalists financed their work differently. Bartram received backing from a wealthy sponsor, while Peale traveled on a government contract, and Audubon self-funded his work through sponsorships. Meanwhile, their work took them in different directions. Bartram explored the Carolinas, Georgia and Florida, Peale journeyed to the region between the Missouri River and the Rockies, and Audubon's sweeping tour took him from Labrador to Florida, Louisiana to the Upper Missouri. These men, however, shared a commitment to contextualizing the natural world they encountered, and portraying the immensity and majesty of what they saw, in order to inspire as well as inform. As the exhibition's title suggests, *Sketching Splendor* documents endeavors that were as much artistic as they were scientific.

But what were revelations to the three men, of course, were not new to the people who lived on the lands they visited. Their sponsors, moreover, cared as much or more about potential profit as scientific discovery. As beautiful and inspiring as the expeditions' images were—and remain—they must also, in other words, be understood as tools for expansion. The careful renderings of Indigenous communities and their circumstances also documented how land was used and its potential for settlement and cultivation. As such, the stunning paintings in this exhibition functioned as advertisements intended to spark interest in regions to the west and south and encourage future development there. Peale, in fact, had been hired primarily for this purpose.

The APS exhibition serves two purposes. Collectively, the objects chart how early natural scientists shifted between efforts to remain dispassionate and objective in the pursuit of their studies, while acknowledging in their art and documentation the wondrous qualities of the phenomena they encountered. Just as important, however, are our careful curatorial efforts to recognize the perspectives and artistic contributions of the women, enslaved people, and Indigenous communities who were integral to these endeavors. Uncovering these perspectives is difficult since they were not central to the purpose of these expeditions. Hearing these often silenced or overlooked voices, however, provides us with a salutary reminder that what you see here today is the work of much more than three sets of scientists' hands.

The American Philosophical Society was founded in 1743 "to promote useful knowledge" by bringing together the greatest minds of the British colonies. Today we interpret that mission more broadly and support it with elected memberships that recognize important scientific and humanitarian contributions, academic conferences, grants, and fellowships to encourage ongoing research, and public engagement through lectures, education programs, and exhibitions like *Sketching Splendor*.

Michelle Craig McDonald
LIBRARIAN / DIRECTOR OF THE LIBRARY & MUSEUM
American Philosophical Society
April 2024

Acknowledgments

Exhibitions and their accompanying catalogs are collaborative efforts made possible by the contributions of many. *Sketching Splendor: American Natural History, 1750–1850* was supported by the dedicated work of the American Philosophical Society staff and contractors, content experts, lenders to the exhibition as well as those who fund our work. With gratitude, we acknowledge the exceptional scholarship of Dr. Anna Majeski, APS Exhibition Researcher and Content Creator, who enthusiastically curated this exhibition and wrote the compelling essay in this catalog. She was a thoughtful and engaging collaborator throughout the exhibition and catalog development.

The dedication and collective efforts of our whole APS Museum Department made this exhibition possible. Associate Director of the Museum Mary Grace Wahl oversaw exhibition development and installation, catalog publication, and managed the exhibitions team with skill and humor. Curatorial Associate Magdalena Hoot made essential contributions to the exhibition's development and installation. Visitor Services Coordinator Deanna Johnson worked with the museum guides and gave feedback on educational moments. Museum Education Manager Ali Rospond and Head of Education Cathy Person provided valued feedback on object labels, developed meaningful educational programming, and prepared our guides with essential training. Curatorial Assistant Nandini Subramaniam facilitated many key administrative and installation components. Lastly, we thank the 2024 Museum Guides for their work in interpreting and sharing the exhibition with the public.

Special thanks are due to our exhibition Advisory Group, Kathleen Foster, Sue Johnson, and Robert M. Peck. Their collective knowledge, feedback, and enthusiastic support were essential to the exhibition's development. Alan Balicki is responsible for graciously providing special assistance with the Audubon watercolors at the New York Historical Society. We thank former APS fellows Bennett Jones, Keith Pluymers, and Gustave Lester for discussions and sharing resources relevant to the exhibition's content. Additionally, we thank Tanya Kevorkian, Debra J. Lindsay, Therese O'Malley, and Amy Meyers, who shared resources and engaged in thoughtful discussions throughout the process. The University of Michigan Library team—Marieka Kaye, Juli McLoone, Caitlin

Pollock, and Jason Young—shared their research on Audubon's *Birds of America*. Amy Meyers, Alexis Wang, and Lora Webb read drafts of the catalog essay.

We also thank the lenders to the exhibition, including The Academy of Natural Sciences at Drexel University, The Historical Society of Pennsylvania, The John James Audubon State Park, and exhibition advisor Robert M. Peck. At the Academy of Natural Sciences, we thank David Cooper, Jessica Lydon, and Nate Rice. At the Historical Society of Pennsylvania, we wish to thank David Brigham and Tara O'Brien, and an extra special thanks to Steve Smith for assisting with the Bartram Collection. At The John James Audubon State Park, we are very grateful to Jennifer Spence and Connor Humphrey. Lastly, we appreciate artists Nadia Hironaka and Matthew Suib, whose video installation, *Field Companion*, added a contemporary perspective to the exhibition content.

We relied on talented contractors during the production of the exhibition and catalog. Barb Barnett deserves praise for her brilliant work handling all the graphic design for the exhibition and catalog. We also thank Exhibition Designer Keith Ragone for redesigning our exhibition layout. Our preparator, Preston Link, makes the exhibition installations a beautiful and functional reality. We rely on several dedicated and skilled professionals, including framer Lucía Torner, mount maker Will Bucher, painters Jeremy and Tammy Morris, copy editor Dan Cook, and Contour Woodworks. Greenhouse Media, consisting of Aaron Igler and Matt Suib, created a compelling Audubon interactive that will live on past the exhibition's close. We also want to thank photographer Brent Wahl for his repeated work on the Audubon folios and other collection items, installation photography, and facsimile production.

The exhibition staff could only produce their exhibitions with the knowledge and support of many other people at the APS. In our Conservation Department, we would like to thank Anne Downey and Renee Wolcott for their tireless efforts in conserving objects in the exhibition, particularly *Birds of America*. We are grateful to the staff of the APS Library for their generous assistance in the reading room and navigating the APS archives. Special thanks to Brian Carpenter for his assistance regarding Indigenous materials and David Gary and Valerie-Ann Lutz for sharing their wealth of knowledge across the APS Library's collections. Brenna Holland read prospectus drafts and object labels and provided valued advice, and Adrianna Link facilitated connections with former and current APS fellows. Our Bryn Mawr summer interns, Spencer Auerbach and Lauren McCouch, contributed to the early research process. Jessica Frankenfield managed publicity for the exhibition, and Linda Jacobs made our work possible with the management of our grants. We also thank our facilities staff, especially Jeremy Schoenrock, for lighting the exhibition and for his willingness to always help out. Finally, thanks to our Executive Officer, Robert M. Hauser, for his unwavering encouragement and support for our work.

The *Sketching Splendor: American Natural History, 1750–1850* exhibition and catalog were made possible by the support of the Mellon Foundation, Pennsylvania Historical and Museum Commission, APS Members and Friends, and donations by visitors to the APS Museum.

Michelle Craig McDonald
LIBRARIAN / DIRECTOR OF THE LIBRARY & MUSEUM
American Philosophical Society
April 2024

Sketching Splendor

AMERICAN NATURAL HISTORY, *1750—1850* [1]

Anna Majeski

Standing before the green expanse of the great Alachua Savanna in Florida, the American naturalist William Bartram (1739–1823) exclaimed, "how is the mind agitated and bewildered, at being thus, as it were, placed on the borders of a new world!" [2] The passage is part of Bartram's *Travels* (1791), recounting his 1773–1777 journey through the American Southeast. Bartram's rapturous declaration punctuates his attempt to enumerate the savanna's seemingly boundless biodiversity and ecological complexity. [3] Trained in the taxonomic tradition of describing individual species, the savanna offered Bartram a more holistic, environmental point of view. [4] What he characterizes as the savanna's "amazing display of the wisdom and power of the supreme author of nature" ultimately suspends human reason and moves him to contemplate nature's sublimity. [5] The passage is justly celebrated by Bartram scholars: taxonomic to ecological, scientific to sublime, Bartram's words create new epistemological worlds. Yet the focus on this passage also elides other, darker ways in which Bartram saw the savanna. [6] The alleged "newness" of a landscape Bartram knows is occupied by the Seminole is also a call-back to colonialist modes of surveillance in natural history writing, in which supposedly vacant and fruitful landscapes offer themselves up for conquest. [7] Philosophical speculation and colonialist interests are bound up inextricably in Bartram's *Travels*, often in quite complex ways.

This exhibition examines three American natural historians—Bartram, Titian Ramsay Peale (1799–1885), and John James Audubon (1785–1851)—whose work exemplifies innovation and injustice in equal measure. The careers of these three naturalists spanned the exciting period from 1750 to 1850, when American science was in its infancy, and the intellectual and creative possibilities in natural history were rich. Yet, this was equally a time when naturalists helped advance expansionist and racist agendas. These are the two realities that this exhibition engages, teasing out the many ways in which natural history relied on and promoted the forces of colonization and plantation slavery.

For Bartram, Peale, and Audubon, the moment was indeed filled with new intellectual opportunities. The writings of European contemporaries provided multiple natural historical models. One approach, exemplified by Swedish botanist Carl Linnaeus (APS, 1769) focused on identifying, naming, and cataloging new species. Linnaeus developed a highly influential and comprehensive taxonomic system that classified plants according to their reproductive parts. On the other hand, French naturalist Georges-Louis Leclerc de Buffon (APS, 1768) thought Linnaeus' system was too simplistic, and instead advocated detailed observation of living things in nature, including their behavior and environmental relations.[8] He also hypothesized natural laws relating to ecology and evolution.[9] Bartram, Peale, and Audubon engaged all these ideas, while also defining American science's distinctive emphasis on empiricism and firsthand experience.[10] Finally, the period covered in this exhibit falls just before the sciences were fully professionalized, when there were fewer fixed rules about natural history as a discipline and genre, and naturalists experimented with ways of understanding nature drawn from art and literature.[11]

Yet natural history was not an innocent intellectual pursuit. The work of Bartram, Peale, and Audubon helped advance colonialist agendas dispossessing Native nations. Annexed lands in turn became sites for the horrors of human chattel slavery. Moreover, natural historians sometimes promoted regimes of knowledge that upheld racial hierarchies, helping to justify the dehumanization of Native Americans and people of African descent.[12] The naturalists in this exhibition also enslaved human beings or benefitted from enslaved labor.[13] As recent public conversations around Audubon have made clear, it is imperative that museum exhibitions engage with both halves of American natural history.[14] At the same time, while Bartram, Peale, and Audubon all supported empire and slavery in various ways, they also differed in the degree to which they embraced these systems of oppression. Bartram, in particular, was intensely ambivalent regarding the dispossession of Native American nations and often publicly defended Native rights. A Quaker, Bartram also changed his views on slavery and racial hierarchies over the course of his life, though his positions remained problematic and far from clear-cut. These changes show him responding to shifting public opinion among his fellow Quakers and the broader American public. Bartram's views often diverge from mainstream biases in his own time.

The worlds shaped by European empires in North America were hierarchical. They were also pluralistic and multicultural. Many peoples of African descent both enslaved and free, as well as Native Americans and women, worked within these conditions to contribute to the making of natural history.[15] Though these contributors were often unnamed, this exhibition has been able to highlight the work of specific individuals whose expert knowledge, skills, and labor helped produce Bartram, Peale, and Audubon's work.[16]

This is the first American Philosophical Society exhibition devoted exclusively to Bartram, Peale, and Audubon, whose work forms a significant part of the Society's natural history collections. It elucidates their different visions of nature in a time of enormous intellectual, social, and political change. Much of the exhibition's content is focused on images, a reflection of the centrality of both verbal *and* visual picturing in the work of natural history, perhaps particularly true of the naturalists in this exhibition.[17] Images also provide us with a unique invitation to take on the full legacy of Bartram, Peale, and Audubon, since they differ in important ways from texts. This essay will touch on three specific phenomena. Firstly, the materiality of the image makes it especially powerful as an ideological vehicle. Images make tangibly real the imagined landscapes of expansionist desires and the cultural misconceptions that reinforce colonialist hierarchies.[18] Imagery is perhaps particularly potent when it substituted for Euro-Americans' firsthand experience of unfamiliar geographies or societies. Secondly, the image intensifies the tensions evident in some natural history texts—particularly Bartram's—between aesthetic reverie or reverence for nature, and the language of expansionism.[19] In a narrative text, the sequential presentation of passages exemplifying these two different attitudes toward nature makes them easier to dissect and treat separately, privileging one or the other; in contrast, the simultaneity of the image intensifies this dialectical tension in ways that are instructive and productive. Finally, the books and illustrations in this show often reflect the racial and gender hierarchies of their contemporary society, as evidenced by whose contributions they do or do not commemorate. However, the formal hierarchies of images—what information is prioritized compositionally—sometimes differs intriguingly from the hierarchies imposed by text-based attributions, which often devalue all but the most elite scientific contributors. Images can offer a starting point for rethinking whose labor counted in the world of American natural history.

William Bartram (APS, 1768) witnessed the emergence of both the American nation and its scientific community. His father, John Bartram, ran an important nursery outside Philadelphia that was also an experimental garden, particularly in the arena of plant ecology.[20] John Bartram lived during a time when European connections were key to colonial science. He tapped into a transatlantic network of merchants, physicians, and naturalists—many of them Quakers, like the Bartrams—and British Quaker patrons like Peter Collinson proved a critical source of support for both John and William Bartram. John was also one of the American Philosophical Society's founding members in 1743, helping to nurture an independent intellectual community in British North America.[21]

William's own career bridged the colonial and early republican periods. His major published work, the *Travels* (1791), surveys a 1773–1777 journey through the Carolinas, Georgia, and East and West Florida, sponsored by the British physician John Fothergill. It is a multifarious work, combining genres in ways that stretched the limits of contemporary natural history writing, and that evidence a complex attitude toward the landscape.[22] Nature is alternately presented as something to be taxonomically cataloged, utilized, and exploited, or else, its 'unimproved' vastness and sublimity is seen as a source of spiritual inspiration. The book's political positionality is equally difficult to nail down, and scholars have viewed it both as an anti-imperial defense of ecological diversity and Native American rights, and as an apology—albeit ambivalent—for the republic's territorial ambitions.[23] In part, the *Travels'* hybrid nature is the product of its long genesis. Between Bartram's departure in 1773 and the book's publication in 1791, he went from loyal British subject advancing the Crown's interests in the American Southeast, to a skeptic of expansionist politics, to a committed citizen of the United States.[24] These shifting attitudes are legible in differences between Bartram's draft manuscript from the 1780s and the published text. The *Travels'* ambivalence is thus a product of Bartram's biography, but it is also part of its potency: condensing many views into a single narrative that reflects the United States' conflicted beginnings.[25]

Bartram's own view probably lay somewhere in the middle of all these positions. Though Bartram was not an avid advocate of American expansion, neither did he oppose it absolutely. Amid the establishment of the United States, he may have believed it would be possible to transform and 'improve' landscapes in ways that were less exploitative and destructive.[26] What such a solution might be, and how best to achieve a balance between competing claims on land in the new nation, he never clarified. Thus, as readers of the *Travels,* we are left with a profoundly unresolved series of questions.

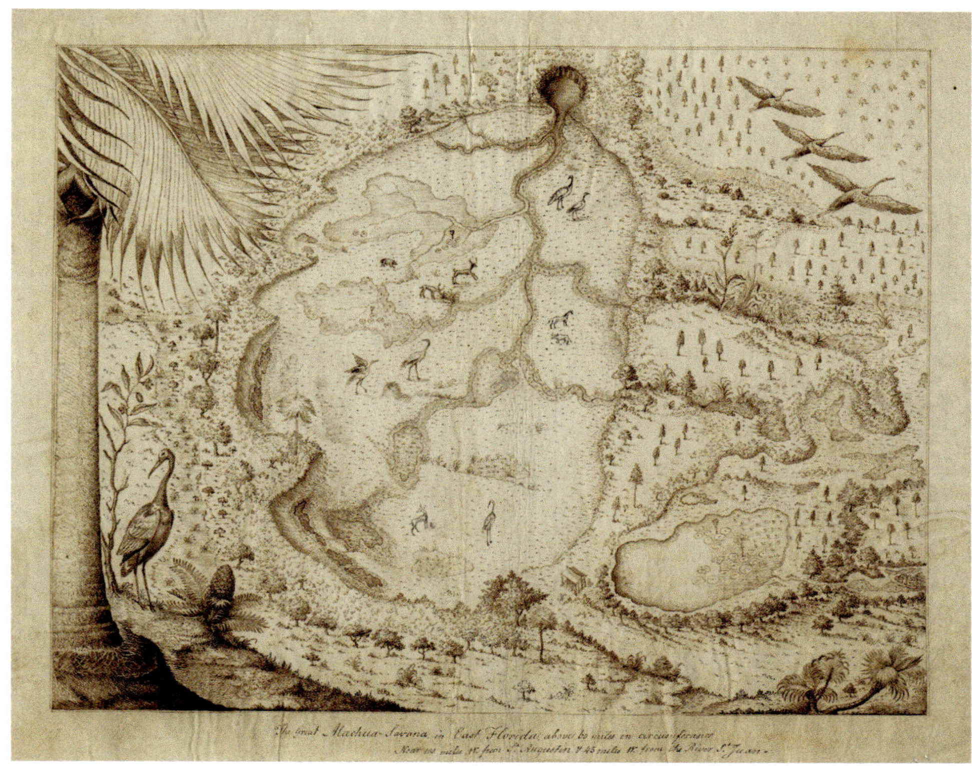

If the *Travels* condenses this ambivalence into a 500-page book, the American Philosophical Society's *The Great Alachua-Savana in East Florida* condenses it into a single image (Figure 1). The savanna, now called Payne's Prairie, bordered the Seminole town of Cuscowilla, represented by the lodge in the bottom right of the APS drawing. Like the *Travels*, the drawing combines representational modes, each signifying different attitudes to the savanna, whether as a source of inspiration or economic enrichment.[27] The high outcropping crowned by a towering palm in the left foreground is rendered in a naturalistic idiom and recalls similar framing devices in picturesque landscape imagery.[28] The outcropping grounds the viewer in the scene, who anticipates a revelation of pleasing contrasts, in which the towering lookout will give way to an attractive view toward the wide, flat savanna, extending back in orderly fashion into space. Instead, the savanna moves abruptly to the vertical, suggesting it is a space that defies perspectival management.[29] This defiance of rational expectation and the laws of linear perspective aligns with Bartram's invocation of the natural sublime in his description in *Travels*, where its vastness and ecological complexity suspends thought and fills him with awe for the greatness of the divine creator.[30] This is language that implies the surrender rather than extension of human control, and an appreciation for the natural world as Bartram found it, in its 'unimproved' state.[31] Yet, the image also suggests cartographical surveillance, and the right half of the APS drawing in particular lacks many of the disorienting

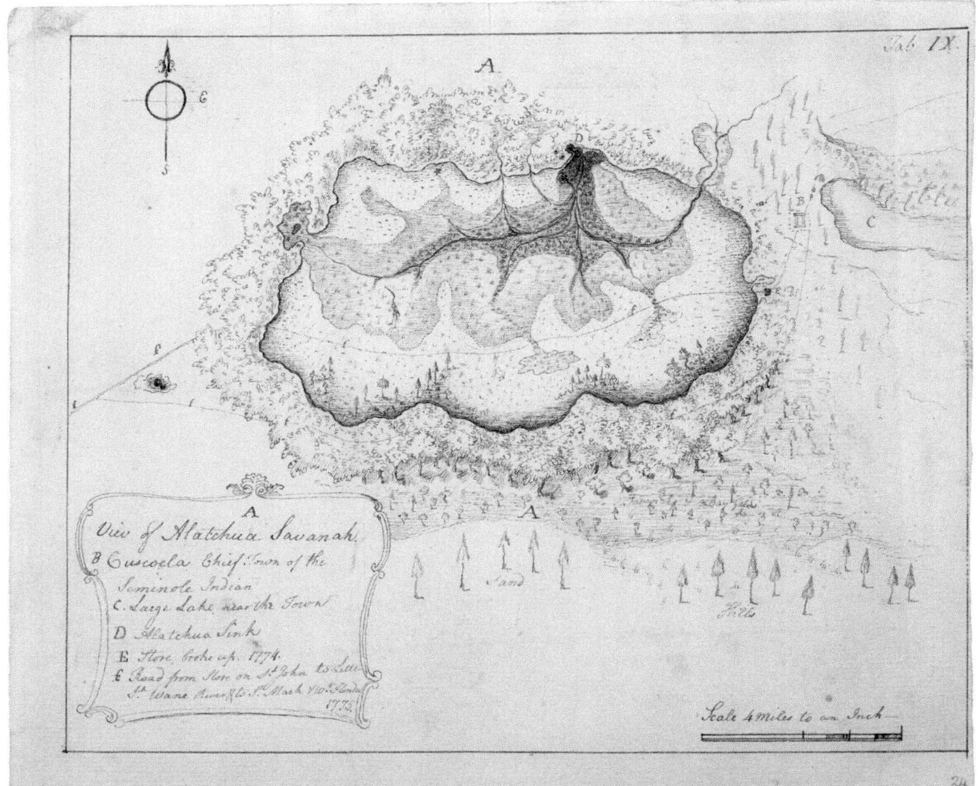

FIG 2
THE GREAT ALATCHUA SAVANAH IN THE PROVINCE OF E[AS]T FLORIDA
William Bartram

1775
Ink on paper
The Trustees of the Natural History Museum, London.

effects of the left.[32] The drawing possesses a map-like elevated view and two-dimensionality, using schematized icons to indicate bodies of water, forests, and animal populations, as well as flattening topographic elements (like the cliffs) into territorial boundary lines.[33] These features also appear in Bartram's other image of the savanna, a map sent to his patron Fothergill in 1775, which offers totalizing surveillance from above, mathematically rendered, complete with compass and scale bar (Figure 2).[34] The APS drawing therefore interweaves picturesque, sublime, and cartographical elements.

The multiplicity of pictorial modes in Bartram's APS drawing mirrors his textual description of the savanna in the *Travels*.[35] Indeed, following Bartram's initial evocation of the natural sublime, he proceeds into a more conventional assessment of the savanna's flora, fauna, and terrain, part of a systematic survey of the site conducted by Bartram and the group of traders with whom he traveled.[36] However, while the same play across categories of language is found in the sequential narrative of the *Travels*, it is intensified in the simultaneous space of the image. Importantly, the various visual modes in the APS drawing are not resolved into any kind of formally cohesive composition. Instead, they remain like fragments of a visual and conceptual collage, creating a dialectical tension between these different ways of understanding nature.[37] *The Great Alachua-Savana* is a pictorial space that invites questions about how we relate to the natural world, and perhaps asks us to find a balance between them. But it does not provide answers.

William also struggled to articulate a position on the competing claims to the landscape, specifically, the claims of Native nations versus the colonialist incursions of the United States government. In the introduction to the *Travels*, Bartram states that one of his goals is to counter the prejudices of white Americans, and to help devise a solution for a peaceful union with Native Americans nations. In parts three and four of the text, he offers a strong and detailed defense of the equality of Native Americans. He praises the dignity of Native American culture at length and its equality with European lifeways, and he defends Native rights to their lands and self-determination.[38] On the other hand, in the *Travels* Bartram also evades the question of how Native American rights will be maintained alongside the American expansion he also anticipates, sometimes positively.[39] For example, despite the time Bartram spent in Cuscowilla, the Seminole town bordering the Alachua Savanna, he anticipates the savanna's appropriateness for a very large, European-style settlement—noting that at least 100,000 people and millions of domestic animals could happily live there.[40] It is clear such a large settlement could not also maintain Seminole sovereignty. Indeed, Bartram's speculation about the site's future probably reflects his awareness of the plans of a planter, Jonathan Bryan, to steal the land through a 99-yr 'lease.'[41] Bartram's position on these conflicts of interest therefore remained unresolved. Ultimately, Bartram would come to support the United States government's 'civilization' policy, which sought to impose European economic, social, religious, and political culture on Native American communities. He wrote to Henry Knox, George Washington's secretary of war, in support of the proposal around 1789.[42] Bartram may have viewed this as a necessary measure to avoid war. Nonetheless, he proved ignorant of the immense violence occasioning government attempts to forcibly erase the cultures of Native American communities. These attempts failed, but not without causing immeasurable harm.

Bartram's address to Knox in support of the government's assimilation policy also offered tacit approval to its plan to promote plantation slavery in Native territories.[43] Paradoxically, the letter to Knox is roughly contemporaneous with Bartram's authorship of an antislavery treatise urging the United States government to reject slavery.[44] Indeed, Bartram's positions on slavery are highly problematic and also changed throughout his lifetime.[45] As a young man in the 1760s, he spent five years on the large North Carolina plantation of his uncle, his father's half-brother, who was also called William Bartram. William Bartram (the naturalist) also ran a plantation in Florida in 1766, where he enslaved six people.

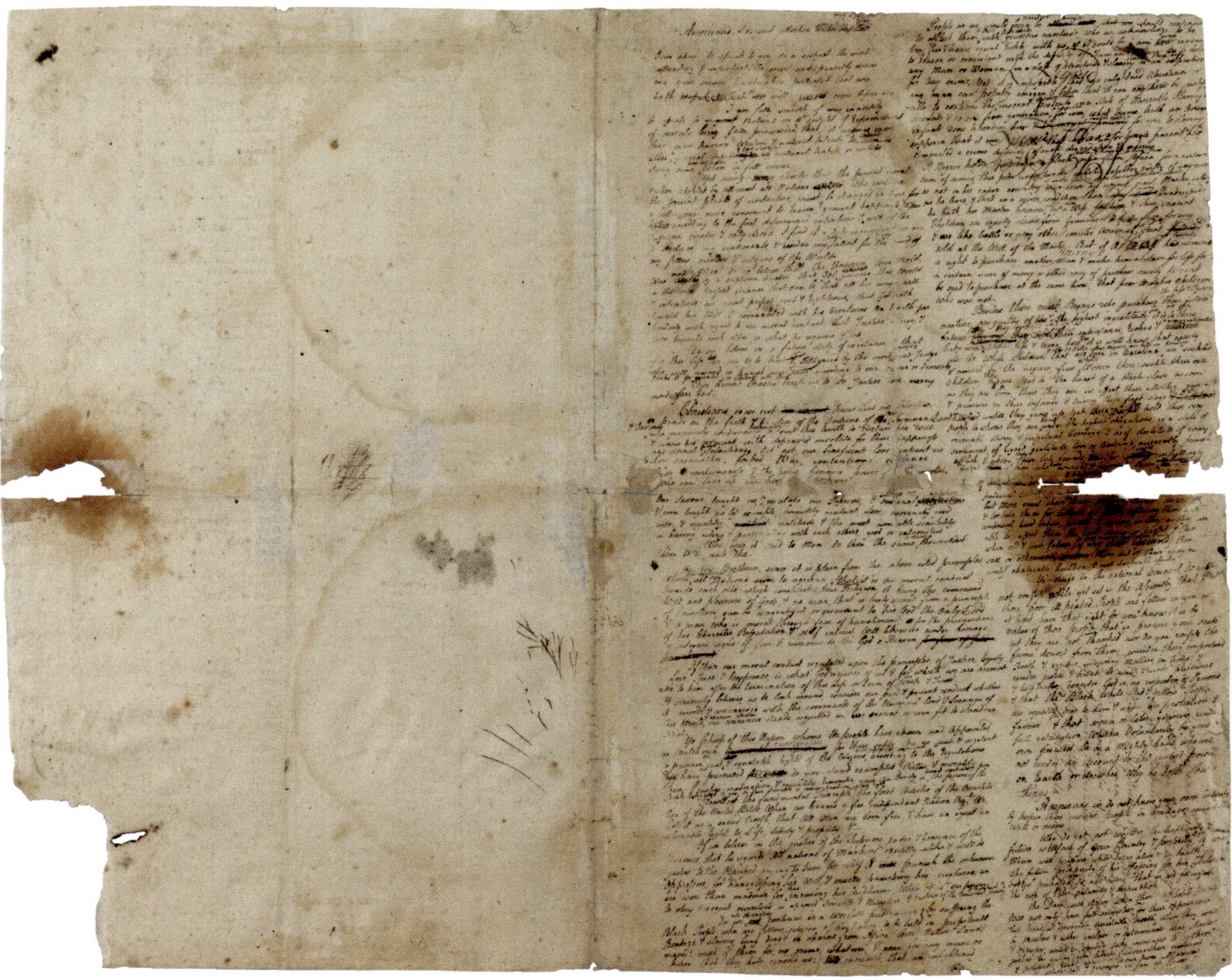

Later, between 1772 and 1773, Bartram enslaved a woman called Jenny, and money from her sale in 1773 may have partially funded his southeastern travels.[46] Perhaps it comes as no surprise that he consistently evades the topic in the *Travels,* even as he revisits the site of his own former plantation in East Florida.[47] On his return to Philadelphia in 1777, William Bartram was greeted by changing public opinion amongst Philadelphia Quakers, who had made enslaving human beings grounds for disownment in 1776.[48] At this point, Bartram seems to have gradually adopted a pro-abolitionist viewpoint; though as his support of the government assimilation policy suggests, this new position was not itself without complexity. Indeed, Bartram's antislavery treatise is a mirror of these quagmires: whether intentionally or by happenstance, William wrote the treatise on the back of a broadsheet catalog of plants for sale in Bartram's Garden (Figure 3a–b).[49] This thriving nursery business implicated the Bartrams in plantation slavery, among

FIG 3A
ANTISLAVERY TREATISE
William Bartram

c. 1790
Written on John Bartram Jr. and William Bartram, *Catalogue of . . . Plants . . . in John Bartram's Garden . . .* (c. 1783)
Broadsheet with manuscript notes
The Collection of the Historical Society of Pennsylvania.

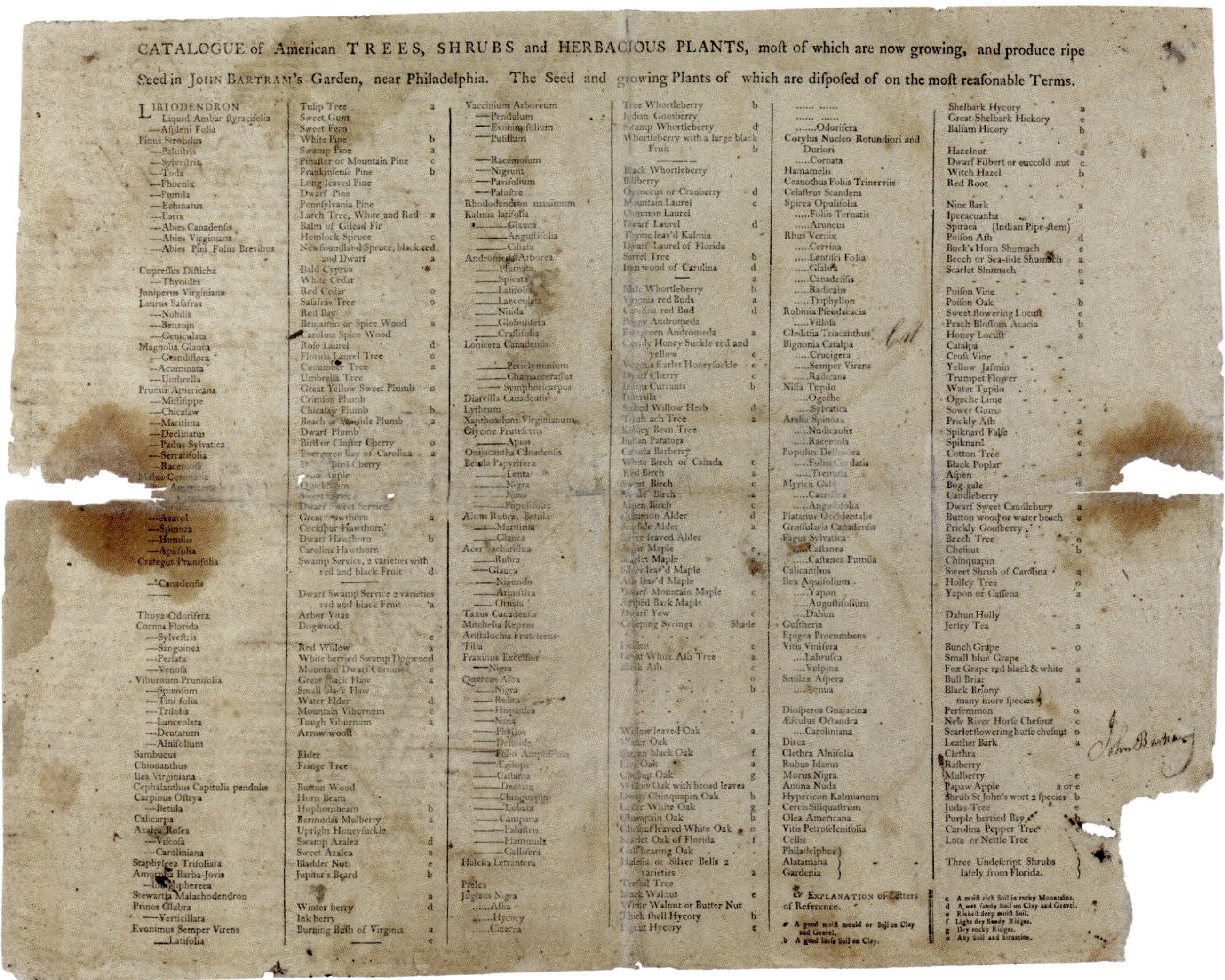

CATALOGUE of American TREES, SHRUBS and HERBACIOUS PLANTS, most of which are now growing, and produce ripe Seed in John Bartram's Garden, near Philadelphia. The Seed and growing Plants of which are disposed of on the most reasonable Terms.

LIRIODENDRON	Tulip Tree
Liquid Ambar styracifolia	Sweet Gum
——Aspleni Folia	Sweet Fern
Pinus Strobilus	White Pine
——Palustris	Swamp Pine — b
——Sylvestris	Pinaster or Mountain Pine — c
——Toda	Frankinsense Pine — b
——Phoenix	Long leaved Pine
——Pumila	Dwarf Pine
——Echinatus	Pennsylvania Pine
——Larix	Larch Tree, White and Red — a
——Abies Canadensis	Balm of Gilead Fir
——Abies Virginiana	Hemlock Spruce — c
——Abies Pini Folus Brevibus	Newfoundland Spruce, black red and Dwarf — a
Cupressus Disticha	Bald Cypres
——Thynoides	White Cedar
Juniperus Virginiana	Red Cedar — o
Laurus Safafras	Safafras Tree — o
——Nobilis	Red Bay
——Benzoin	Benjamin or Spice Wood — a
——Geniculata	Carolina Spice Wood
Magnolia Glauca	Rose Laurel — d
——Grandiflora	Florida Laurel Tree — o
——Acuminata	Cucumber Tree — o
——Umbrella	Umbrella Tree
Prunus Americana	Great Yellow Sweet Plumb — o
——Mississippe	Crimson Plumb
——Chicataw	Chicataw Plumb — b
——Maritima	Beach or Seaside Plumb — a
——Declinatus	Dwarf Plumb
——Padus Sylvatica	Bird or Chuler Cherry — o
——Serratifolia	Evergreen Bay of Carolina — a
——Racemosa	Bird Cherry
Malus Coronaria	——Lenta
——Azarol	Quicksigum
——Spinosa	Sweet Service
——Ramulus	Dwarf Sweet Service
——Apifolia	Great Hawthorn
Crategus Prunifolia	Cockspur Hawthorn
	Dwarf Hawthorn — b
——Canadensis	Carolina Hawthorn
	Swamp Service, 2 varieties with red and black Fruit — d
	Dwarf Swamp Service 2 varieties red and black Fruit — a
Thuya Odorifera	Arbor Vitae
Cornus Florida	Dogwood — e
——Sylvestris	Red Willow
——Sanguinea	White berried Swamp Dogwood
——Perlata	Mountain Dwarf Cornus — d
——Venosa	Great black Haw — e
Viburnum Prunifolia	Small black Haw — a
——Spinosum	Water Elder — d
——Tinifolia	Mountain Viburnum — e
——Triloba	Tough Viburnum — a
——Lanceolata	Arrow wooll
——Dentatum	
——Alnifolium	
Sambucus	Elder — e
Chionanthus	Fringe Tree
Ilea Virginiana	
Cephalanthus Capitula pendula	Button Wood
Carpinus Ostrya	Horn Beam
——Betula	Hophornbeam — b
Calicarpa	Bermudas Mulberry — a
Azalea Rosea	Upright Honeysuckle
——Viscosa	Swamp Azalea — d
——Caroliniana	Sweet Azalea — a
Staphylea Trifoliata	Bladder Nut — e
Amorpha Barba-Jovis	Jupiter's Beard
——Spherea	
Stewartia Malacodendron	
Prinos Glabra	Winter berry — d
——Verticillata	Ink berry
Evonimus Semper Virens	Burning Bush of Virginia — a
——Latifolia	

Vaccinium Arboreum	Tree Whortleberry — b
——Pendulum	Indian Goosberry
——Evonimifolium	Swamp Whortleberry — d
——Putisium	Whortleberry with a large black Fruit — b
——Racemosum	Black Whortleberry
——Nigrum	Bilberry
——Parvifolium	Oxicoccus or Cranberry — d
——Palustre	Mountain Laurel — c
Rhododendron maximum	Common Laurel
Kalmia latifolia	Dwarf Laurel — d
——Glauca	Thyme leav'd Kalmia
——Angustifolia	Dwarf Laurel of Florida
——Ciliata	Sorrel Tree — b
Andromeda Arborea	Iron wood of Carolina — d
——Plumata	
——Spicata	Male Whortleberry — b
——Latifolia	Virginia red Buds — a
——Lanceolata	Carolina red Bud — d
——Nitida	Boggy Andromeda
——Globulifera	Evergreen Andromeda — a
——Crassifolia	Cloudy Honey Suckle red and yellow — e
Lonicera Canadensis	Virginia scarlet Honeysuckle — e
——Periclymenum	Dwarf Cherry — c
——Chamaecerasus	Indian Currants — b
——Symphoricarpos	Diarvilla
Diarvilla Canadensis	Spiked Willow Herb — d
Lythrum	Tooth ach Tree — a
Xanthoxilum Virginianum	Kidney Bean Tree
Glycine Frutescens	Indian Patatoes
——Apios	Canada Barberry
Oxiacantha Canadensis	White Birch of Canada — e
Betula Papyrifera	Red Birch — a
——Lenta	Sweet Birch — e
——Nigra	Dwarf Birch — a
——Nana	Aspen Birch — c
——Populifolia	Common Alder — d
Alnus Rubra, Betula	Sea side Alder — a
——Maritima	Silver leaved Alder
——Glauca	Sugar Maple — e
Acer Sachariflua	Scarlet Maple — a
——Rubra	Silver leav'd Maple — e
——Glauca	Ash leav'd Maple
——Nigundo	Dwarf Mountain Maple — e
——Arbutiva	Striped Bark Maple
——Ornata	Dwarf Yew
Taxus Canadensis	Creeping Spirea — Shade
Mitchelia Repens	
Aristalochia Fruticens	Hidden
Tilia	Linden
Fraxinus Excelsior	Great White Ash Tree — e
——Nigra	Black Ash — o
Quercus Alba	
——Nigra	
——Rubra	
——Hispanica	
——Nana	
——Phyllos	Willow leaved Oak — a
——Deltoide	Water Oak
——Folio Amplissima	Barren black Oak — f
——Agilops	Live Oak — e
——Castana	Chesnut Oak
——Dentata	Willow Oak with broad leaves
——Chinquapin	Dwarf Chinquapin Oak — g
——Lobata	Lesser White Oak — g
——Campana	Champian Oak — o
——Palustris	Chesnut leaved White Oak — d
——Flammula	Scarlet Oak of Florida — f
——Callifera	Gall bearing Oak
Halesia Tetrantera	Halesia or Silver Bells 2 varieties — a
	Trefoil Tree
Ptelea	
Juglans Nigra	Black Walnut
——Alba	White Walnut or Butter Nut — a
——Hycory	Thick shell Hycory — b
——Cinerea	Great Hycory

....	Shelbark Hycory — a
	Great Shelbark Hickory — e
......Odorifera	Balsam Hicory — b
Corylus Nucleo Rotundiori and Duriori	Hazelnut — a
......Cornata	Dwarf Filbert or euccold nut — c
Hamamelis	Witch Hazel — b
Ceanothus Foliis Trinerviis	Red Root
Celastrus Scandens	Nine Bark — a
Spirea Opulifolia	Ipecacuanha
......Foliis Ternatis	Spiraea (Indian Pipe-stem)
......Aruncus	Poison Ash — d
Rhus Vernix	Buck's Horn Shumach — e
......Cervina	Beech or Sea-side Shumach — a
......Lentisci Folia	Scarlet Shumach — o
......Glabra	
......Canadensis	Poison Vine
......Radicans	Poison Oak — b
......Triphyllon	Sweet flowering Locust — e
Robinia Pseudacacia	Peach Blossom Acacia — b
......Villosa	Honey Locust — a
Gleditsia Triacanthus	Catalpa
Bignonia Catalpa	Cross Vine
......Crucigera	Yellow Jasmin
......Semper Virens	Trumpet Flower
......Radicans	Water Tupilo
Nissa Tupilo	Ogeche Lime
......Ogeche	Sower Gum
......Sylvatica	Prickly Ash — a
Aralia Spinoza	Spiknard False — e
......Nudicaulis	Spiknard — e
......Racemosa	Cotton Tree — a
Populus Deltoidea	Black Poplar
......Foliis Cordatis	Aspen
......Tremula	Bog gale — d
Myrica Gale	Candleberry
......Caerulea	Dwarf Sweet Candlebury — a
......Angustifolia	Button wood or water beach — e
Platanus Occidentalis	Prickly Goosberry
Grossularia Canadensis	Beech Tree — o
Fagus Sylvatica	Chesnut — b
......Castanea	Chinquapin
......Castanea Pumila	Sweet Shrub of Carolina — a
Calicanthus	Holley Tree — o
Ilex Aquifolium	Yapon or Caffena — a
......Yapon	
......Augustifolium	Dahun Holly
......Dahun	Jersey Tea — a
Gaultheria	
Epigea Procumbens	Bunch Grape — o
Vitis Vinifera	Small blue Grape
......Labrusca	Fox Grape red black & white — a
......Vulpina	Bull Briar — a
Smilax Aspera	Black Briony
	many more species
Diospiros Guaiacina	Persemmon — o
Æsculus Octandra	New River Horse Chesnut — e
......Caroliniana	Scarlet flowering horse chesnut — o
Dirca	Leather Bark — a
Clethra Alnifolia	Clethra
Rubus Idaeus	Rasberry
Morus Nigra	Mulberry — e
Anona Nuda	Papaw Apple — a or e
Hypericon Kalmianum	Shrub St John's wort 2 species — b
Cercis Siliquatrum	Judas Tree — e
Olea Americana	Purple berried Bay — b
Vitis Petroselenifolia	Carolina Pepper Tree — e
Celtis	Lote or Nettle Tree
Philadelphus	
Alatamaha	Three Undescript Shrubs
Gardenia	lately from Florida.

EXPLANATION of Letters of Reference.

a A good mould or Soil on Clay and Gravel.
b A good loose Soil on Clay.
c A moist rich Soil in rocky Mountains.
d A wet sandy Soil on Clay and Gravel.
e Richest deep moist Soil.
f Light dry Sandy Ridges.
g Dry rocky Ridges.
o Any Soil and Situation.

[manuscript signature:] John Bartram

FIG 3B
VERSO: **CATALOGUE OF … PLANTS … IN BARTRAM'S GARDEN …**
John Bartram Jr.
and William Bartram

c. 1783
Broadsheet with
manuscript notes
The Collection of the
Historical Society of
Pennsylvania.

other things via regular shipments of plants from nurseries in Charleston, where slave-based agriculture was practiced.[50] William probably never came to terms with the garden's relationship to slavery, but instead saw in it a site for potential renewal.[51] The antislavery treatise's double-sidedness offers further parallels with the ambivalence in *The Great Alachua-Savana*, as it emblematizes how William never cohesively resolved his late-life disavowal of slavery with his own implication in that same system.

Bartram's work, in particular *The Great Alachua-Savana*, encapsulates many of the exhibition's themes. In a single image, it speaks to the conflicted nature of early American politics, and the place of natural historians like William Bartram in that complex landscape.

William Bartram was a formative intellectual influence for Titian Ramsay Peale (APS, 1833). Peale knew Bartram from childhood thanks to his father Charles Willson Peale (APS, 1786), a painter by training who founded one of the first American natural history museums, and Bartram informed Titian's commitment to firsthand observation and fieldwork.[52] Titian and his contemporaries viewed themselves self-consciously as realizing a project that Bartram had started, to assemble an ever-more complete catalog of native American flora and fauna.[53] Yet, Titian's natural history also differs in key respects from that of his predecessor. For one, Titian lived during a time when an increasingly professionalized naturalist community emphasized the rigorous adoption of modern taxonomic principles, leaving less room for other ways of describing nature.[54] Moreover, in contrast to Bartram, Titian embraced the prospect of American expansionism with great enthusiasm. Charles Willson Peale was no doubt a critical influence in both these respects. At the Peale Museum, Charles assembled as many American specimens as possible in strict taxonomic order (Figure 4). The collection's growth was fueled by government surveys that aimed to extend the United States' territorial control.[55] In late 1818, Charles secured a place for his young son Titian on an 1819–1820 government-funded survey of the region between the Missouri River and the Rockies, led by Stephen H. Long (APS, 1823).[56] In the official watercolors that Titian executed as assistant naturalist on the Long Expedition, he did not just catalog American nature, he presented it as harmonious, orderly, and open to white settlement.

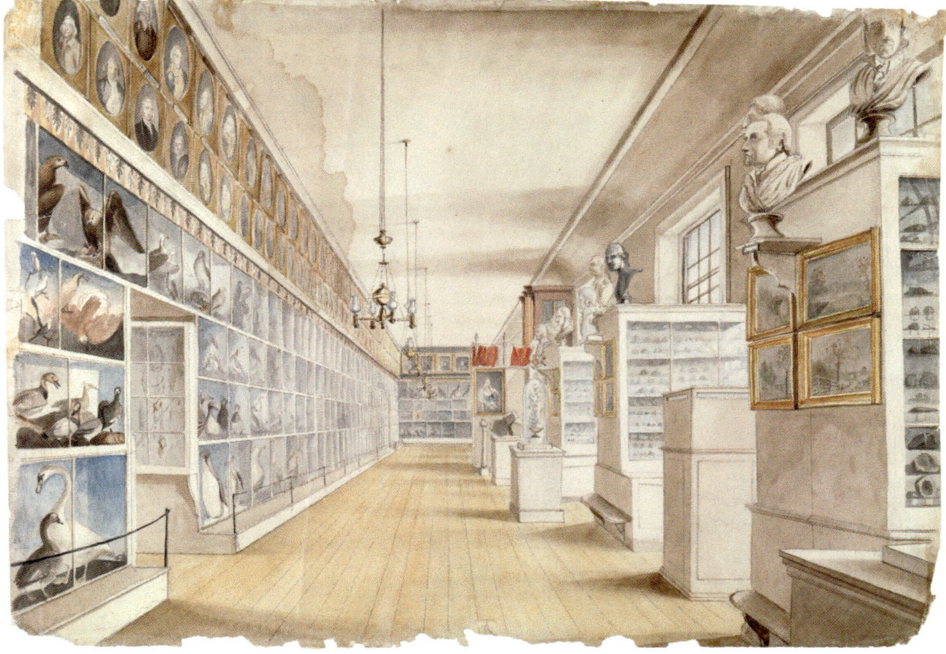

FIG 4
THE LONG ROOM, INTERIOR OF FRONT ROOM IN PEALE'S MUSEUM
Charles Willson Peale and Titian Ramsay Peale

1822
Watercolor over graphite pencil on paper
Detroit Institute of Arts, Founders Society Purchase, Director's Discretionary Fund, 57.261.

FIG 5 (OPPOSITE)
SANDHILL CRANES
Titian Ramsay Peale

1820
Watercolor and graphite on paper
APS.

The Long Expedition was originally conceived as the scientific arm of a military excursion in the Upper Missouri River region.[57] Due to mismanagement and a shortage of government funds, however, the expedition's directives changed, and the military mission failed. After exploring the Upper Missouri River region near Council Bluffs (modern Nebraska) during the fall and winter of 1819–1820, in June 1820 Long led the expedition party west over the Great Plains. Once they reached the front range of the Rocky Mountains, they headed south, and followed the Arkansas and Canadian Rivers back east. Through the Long Expedition, the United States government was particularly anxious to supplement the scientific and visual data gathered during the Lewis and Clark Expedition (1804–1806).[58] Titian Ramsay Peale was one of a number of trained scientists and artists assembled for this task.[59] The official directives given to the expedition members clearly anticipate future territorial expansion: to provide as complete a portrait as possible of the topography, terrain, soil, bodies of water, and interesting animals, plants, and minerals.[60]

Titian's finished watercolors reflect the intertwined goals of intellectual and territorial control. In *Sandhill Cranes* (Figure 5) he confidently asserts human reason's capacity to master natural phenomena.[61] Whereas the conventional

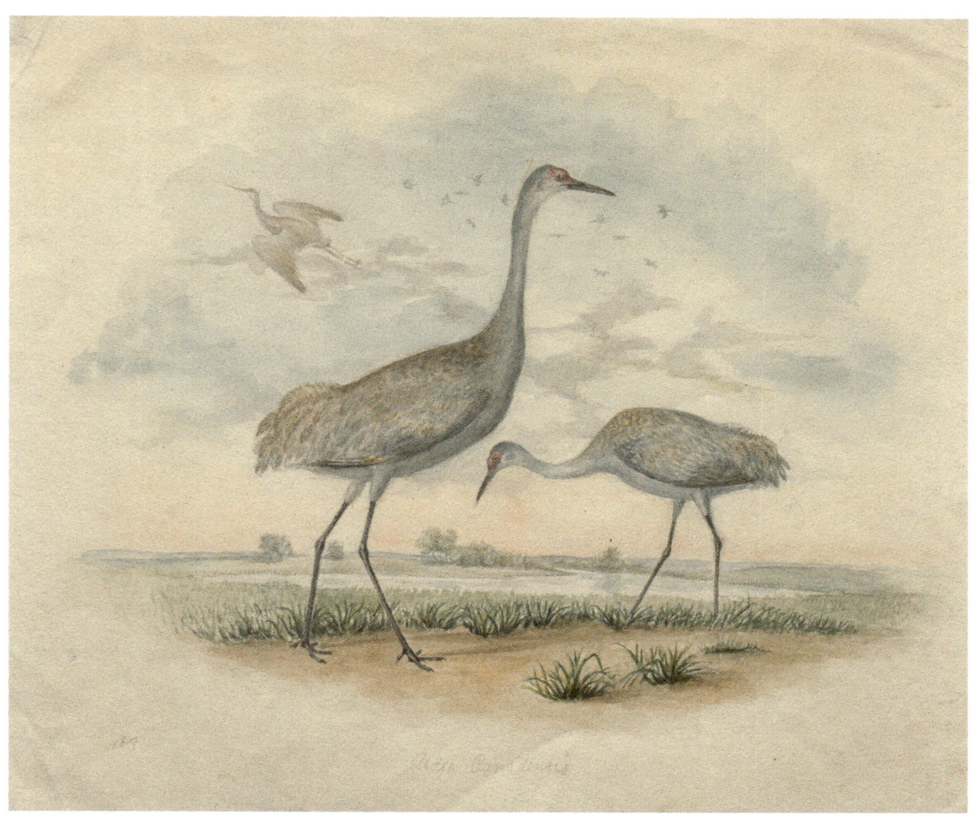

specimen portrait represented only the species' characteristic visual appearance, Peale also summarizes the birds' feeding habits, sociability and tendency to move in large flocks, as well as the species' Platte River wetlands habitat.[62] The image is a testament to Peale's achievement as a field naturalist, in which prolonged empirical observation culminates in an authoritative digest of the species and its life history. Indeed, the composition's insistent circularity—a geometrical figure of self-contained completeness—seems to be a device that proclaims the image's scientific comprehensiveness. Whereas naturalists like Bartram saw animals as beings with an ineffable inner life that surpassed understanding, Peale proclaims they are a field over which humans exercise total intellectual control.[63] Equally significant is the image's overall emphasis on visual order and clear perspectival sightlines.[64] These are surely rhetorical figures for nature's total susceptibility to human reason, but they also enact a fantasy of territorial penetration. From foreground to background, the eye is guided smoothly back, moving from lush grassland to the wide Platte River, to low trees, and finally to a rosy sky, all interlocking in a harmonious composition recalling pastoral landscapes.[65] These are the very natural features that also bespeak the region's promise for future settlement. Peale's presentation of this landscape as orderly, harmonious, and easily graspable signals not just confidence in humanity's intellectual control over nature, but also in the promise of this environment for territorial expansion.[66] These scientific and ideological goals are inextricably bound up together in a single image.

While Peale surely strove for objectivity in his images, he subconsciously shaped the geographical realities he encountered in ways that reflected contemporary cultural preconceptions.[67] In particular, Peale's diaries evidence his investment in the Jeffersonian myth that the lands west of the Mississippi River were a garden of agricultural opportunity that replicated the eastern United States.[68] These preconceptions clearly informed Peale's particular pastoral language, in which these unfamiliar landscapes are remade to reflect European and Euro-American ideals of natural beauty, inextricable from the land's potential fecundity. In Peale's watercolors, art, science, and culture collectively reshape concrete geographical realities, producing imagined landscapes that reflect expansionist desires.

However not all the expedition's members shared Peale's optimistic vision of this territory. In fact, Stephen Long, Edwin James the expedition's botanist and geologist, and William Swift, the assistant topographer, all viewed the lands between the Missouri and the Rockies as arid, harsh, and near uninhabitable.[69] It was this vision of the Great Plains that James would disseminate in his official report on the Long Expedition, *Account of an Expedition from Pittsburgh to the*

Rocky Mountains (1822), where it was reinforced by maps authored by Long and Swift, labeling the entire region a 'Great American Desert.'[70] While the vast majority of Peale's watercolors express an optimistic attitude toward expansion, he must have experienced some ambivalence as two of Peale's watercolors from the APS collections reflect the pessimism found in the report, *Dusky Wolf Devouring Deer Head* and *Magpie* (Figures 6–7).[71] These compositions both depict a harsh and unforgiving environment nearly bare of living

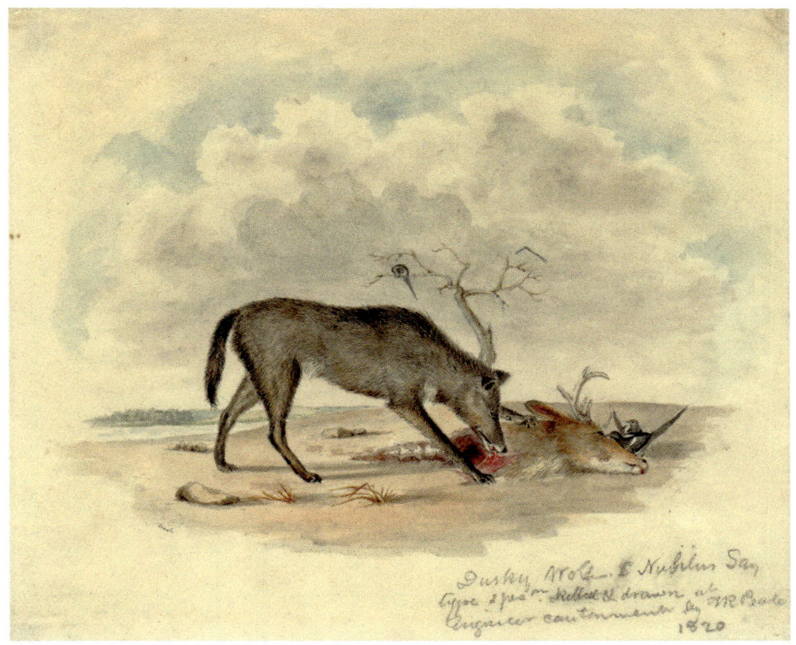

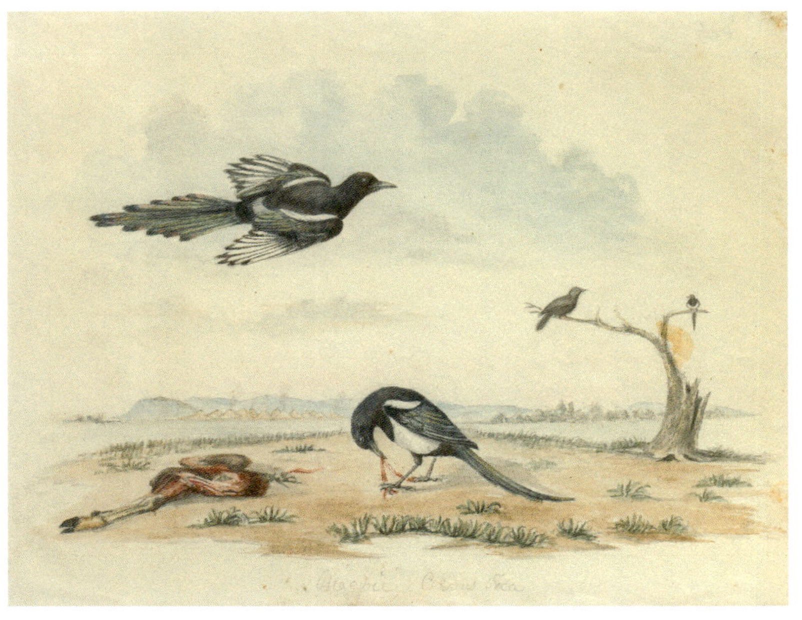

vegetation, in which wolves and birds are left to scavenge dead animals. Instead
of the clear view of the *Sandhill Cranes,* a thick, stormy sky obscures much
of the view behind the *Dusky Wolf,* as if the landscape itself resists visual and
physical exploration.[72] Peale also hints that obstacles to expansion are more than
environmental, representing the tipis of a Native American encampment in
the background of *Magpie* (Figure 7a).[73] The watercolor physically conflates the

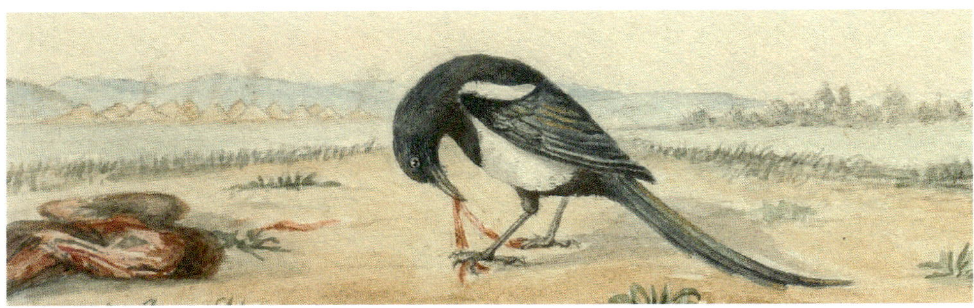

Native American encampment with the natural world, the waving, organic lines
of the tipis inviting visual comparisons with the mountain range beyond and the
tree line on the opposite shore.

Peale's association of human cultures and geography fits the environmental
logic of his natural history studies: just as climate was a means to define and
distinguish flora and fauna, it was also a means to categorize human bodies.
Indeed, during this period, human cultures were typified and categorized
according to climate in ways that justified existing racial hierarchies and colonial
interests.[74] And alongside these ideas about the ways in which climate shaped
human bodies were fears about the potentially malevolent influence of certain
environments on Euro-American bodies. 'Savage' environments and proximity to
non-white peoples threatened those who wanted to create an image of the United
States as a white nation.[75] These patently racist fears surely lay behind Peale's
sketch. Notwithstanding these two watercolors, however, the tenor of Peale's
finished watercolors is largely optimistic about expansion.

Peale's watercolors and James's *Account,* with its maps by Long, reflect two
distinctive visions of the Great Plains: one as garden and the other as desert
waste. Interestingly, none of Peale's natural history watercolors–overwhelmingly
positive notwithstanding the two exceptions discussed— were illustrated in the
published account, though they would inform the Peale Museum displays.[76]
Neither James and Long, nor Peale, adequately describe the huge, ecologically
diverse region that stretched from the Missouri to the Rockies, and the
Yellowstone to the Red and Arkansas Rivers. Both parties offer willful cultural

misperceptions, informed by a complex range of factors and desires. In the case of James and Long's 'Great American Desert,' they tapped into an existing counter-narrative to Jeffersonian optimism, as Zebulon Pike had already described the area in these terms in 1806.[77] But an equally important factor for James was the group's consultation with a number of Pawnee chiefs right before their departure for the Rockies.[78] Long-Hair, of the Chaui Pawnees; Fool Robe, of the Kitkehahki Pawnees; and Knife-Chief, of the Skidi Pawnees, all cast serious doubts on the wisdom of the expedition's endeavor, depicting the territory as devoid of water, game, and timber, and full of enemies.[79] James himself guessed that they might have ulterior motives, and that the Pawnee were in fact seeking to fend off any incursions on important hunting grounds. Nonetheless, it is clear from the *Account* that the expedition members failed to gain important intelligence about their route from the Pawnees. Thus, the Long Expedition set out across the Great Plains at a great disadvantage, and indeed, one group got lost while trying to locate the source of the Red River.[80] As scholars have recently underlined, both sharing *and* withholding information was a key means by which Native American experts asserted their agency and controlled the flow of politically useful information.[81] While James and Long's 'Great American Desert' surely had many origins, the refusal of Pawnee authorities to share critical knowledge no doubt contributed to the expedition's inability to intellectually—and physically—navigate the landscape. If the frustration of American incursions onto the Great Plains was the goal of the Pawnee, they were successful, as the Long Expedition's negative assessment dampened enthusiasm for American expansion in the region.[82]

The different descriptions of the Great Plains that emerged out of the Long Expedition attest to the debates about expansion amongst contemporary Americans. Different parties sought to shape that debate through various media—texts, maps, images, and museum displays. Rather than objective presentations of these geographies, these different media transmitted projections, crafted to respond to divergent ideological desires. Images, with their capacity to offer Euro-American audiences in the east tangible visions of distant landscapes, were a particularly powerful tool. Ultimately, however, the vision of the Great Plains as a desert was not just a construction by white Americans. It also might evidence how the Pawnee actively crafted Euro-American perceptions of the Great Plains in deliberate and strategic ways.

Of the three naturalists included in this exhibition, John James Audubon (APS, 1831) undoubtedly has the greatest name recognition today. Audubon was born in 1785 in the French colony of Saint-Domingue (modern Haiti) to Jean Audubon, a sea captain and plantation owner, and a servant in his father's household who may have been of part-African descent. He was sent to France three years later during the unrest that culminated in the Haitian Revolution, and came to the United States in 1803 where he cultivated an interest in ornithology.[83] Although Audubon was an outsider within the Philadelphia naturalist community compared with Bartram or Peale, from an early age he familiarized himself with the literature of natural history, and he knew Bartram's *Travels*.[84] Audubon's *The Birds of America* (1827–1838) is his best-known project, praised for its dramatic, multi-figure compositions, showing living birds in their ecological contexts, their complex social lives, and their inner emotional reality.[85] The images have become synonymous with the American conservation movement.[86] Recently, however, scholars and the public have collectively rethought Audubon's status as an American icon.[87] Audubon directly contributed to the colonialist and racist agendas discussed in this exhibition. He was an apologist for slavery who enslaved human beings in the 1810s and 1820s, and he opposed abolition.[88] He also promoted scientific racism, collecting the remains of no less than 15 individuals of Mexican, Hispano-Indian, and Native American heritage for the Philadelphia physician and a founding figure of scientific racism, Samuel G. Morton (APS, 1828).[89] Audubon's two major projects, *Birds* and *The Quadrupeds of North America* (1845–1848), likewise promoted these agendas.[90] Both projects solidified Euro-American claims on the landscape at the expense of Native Americans facing intensifying dispossession and romanticized the brutality of human chattel slavery.[91] While legitimate questions persist around Audubon's potentially mixed-race heritage, he consistently sided with slavery and white supremacy both publicly and privately.[92] Audubon's images humanize birds, but he dehumanized peoples of African descent and Native Americans.

One path forward is to ask who else's story we might tell through Audubon's work. Indeed, *Birds* and *Quadrupeds* were collaborative projects involving many people. Audubon's family, including his two sons and his wife Lucy, his printers Richard Lizars and Robert Havell Jr., scientific collaborators such as William MacGillivray and John Bachman, the taxidermist Henry Ward, and many artist-assistants including Joseph Mason and George Lehman, all contributed.[93] Importantly, there were also many unnamed Native Americans and people of African descent who made Audubon's work possible.[94] Our exhibition focused on three individuals in particular: Natoyist-Siksina' (Medicine Snake Woman), a Kainai woman who

helped facilitate the multiracial knowledge networks on which Audubon relied;
Maria Martin, the sister-in-law and later wife of Audubon's collaborator John
Bachman, who produced botanical images for many plates in *Birds*; and a man
named Thomas, whom Bachman enslaved, and who taxidermied specimens during
the years Audubon collaborated with Bachman on *Birds*.[95]

One of the most remarkable objects in the exhibition is Audubon's portrait
of Natoyist-Siksina' with her young child from the collections of John James
Audubon State Park (Figure 8).[96] Natoyist-Siksina' was the daughter of Two Suns,
a Kainai chief (Blackfoot Confederacy), and she married Alexander Culbertson,
an agent of the Chouteau & Co Fur Company, in 1840 (Figure 9).[97] Culbertson
was superintendent at Fort Union when Audubon and his associates visited in
summer 1843 to conduct fieldwork for *Quadrupeds*.[98] Audubon and Natoyist-
Siksina' interacted frequently, but his journal provides few hints about the natural
knowledge she may have shared with him.[99] However Natoyist-Siksina's role
transcended knowledge-sharing. As the daughter of a powerful Kainai chief, her
marriage to Culbertson maintained a delicate alliance between the Blackfoot
Confederacy and the fur company.[100] She was more than a figurehead, and she
and her husband were the most important intermediaries between the Blackfoot
Confederacy and the Chouteau & Co Fur Company on the Upper Missouri.[101]
On Audubon's arrival, Culbertson and his wife offered him the help of their

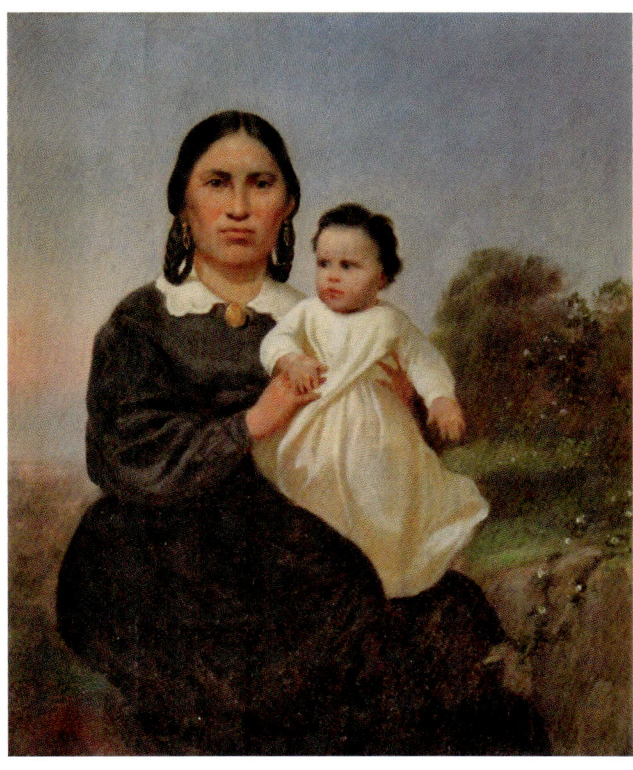

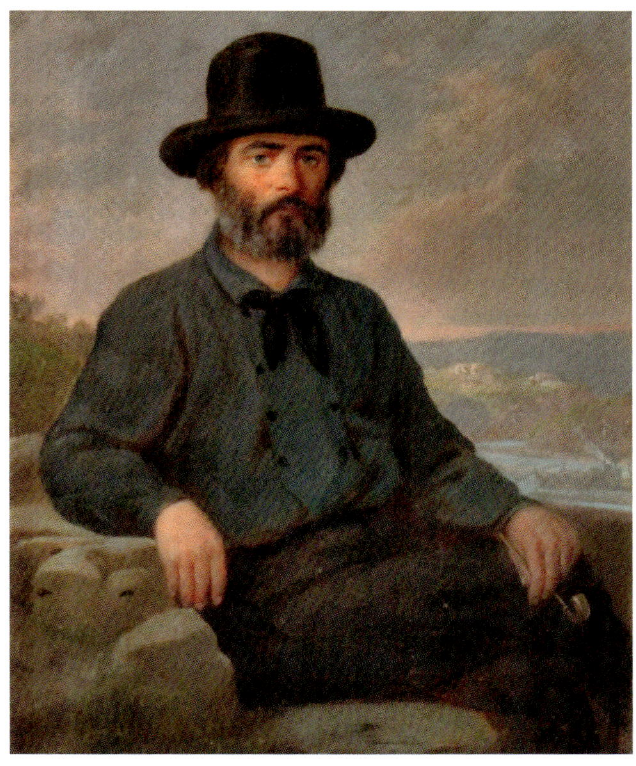

network of Native and white hunters.[102] Audubon came to rely on one man in particular, Owen Mackenzie, of half-Blackfoot heritage, as both a guide and a hunter.[103] Natoyist-Siksina's political and diplomatic negotiations helped to maintain relations between the Blackfoot, other Native nations, and fur company employees, thereby fostering the flow of both information and specimens across permeable cultural boundaries. Audubon directly relied on the cultural space Natoyist-Siksina' helped to construct.

Audubon repaid the graciousness of his two hosts through the gift of a pair of pendant portraits.[104] Audubon represents Natoyist-Siksina' as both beautiful and intelligent, qualities he remarked on in his journal.[105] He has invested particular care in her eyes, with their steadfast and sharp gaze, her dark hair, and well-defined bone structure. Yet the lines between humanizing individuation and dehumanizing physiognomical typecasting are fine and Audubon's ornithological studies had long accustomed him to reading visible qualities as marks of difference.[106] Jet black hair and high cheekbones were both qualities stereotypically associated with Native Americans. Audubon also clearly differentiates the hue of Natoyist-Siksina's complexion from that of her baby, whose skin bears greater resemblance to the child's father. In her European-style dress, Audubon's portrait intentionally invites both identification and distance from Natoyist-Siksina', a tension that Audubon also constructs in his written portrait of her.[107]

The backgrounds of the two portraits, possibly by Audubon's assistant Isaac Sprague, reinforce these disempowering projections onto Natoyist-Siksina's person.[108] The landscape over Natoyist-Siksina's left shoulder is inchoate and hazy, so that her body seems to float, and is anchored only by the rocky boulder on her right connecting her to her husband. To the right of Culbertson's nonchalant form, a steamboat proceeds up the Missouri River (Figure 9a). Audubon presents Natoyist-Siksina' as a woman literally caught between two worlds, one which is receding out of focus, and the other coming into being. The composition is a clear callback to the stereotypical emphasis on change in images of the American West, in which contemporary Native American cultures are disappeared into a primitive and ill-defined past in the face of advancing 'modernity.'[109]

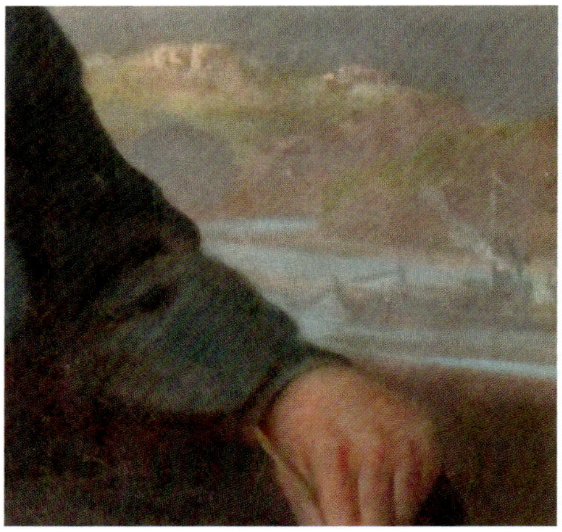

It is important to underline that the narratives in Audubon's portrait ran contrary to the political reality he encountered at Fort Union. In the 1840s, the Kainai and the broader Blackfoot Confederacy were a dominant political force in the

FIG 9A
DETAIL: **ALEXANDER CULBERTSON**
John James Audubon

c. 1843
Oil on canvas
John James Audubon State Park, Kentucky Department of Parks.

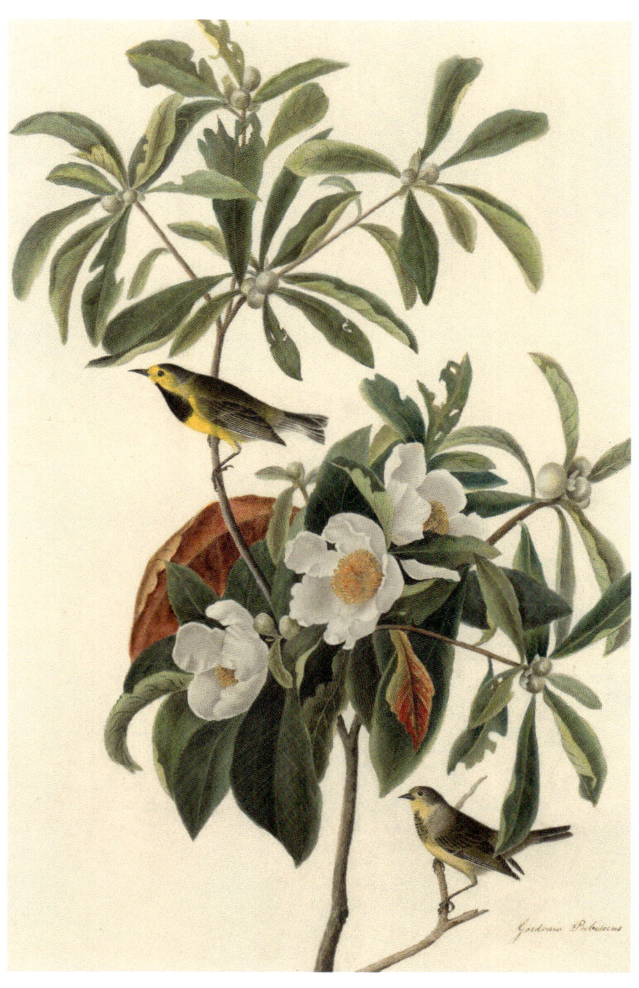

FIG 10
**BACHMAN'S WARBLER
(VERMIVORA BACHMANII),
STUDY FOR HAVELL PL. 185**
John James Audubon and
Maria Martin

1833
Watercolor, graphite, gouache,
and black ink with scratching
out on paper, laid on card.
Mat: 29 × 23 in. (73.7 × 58.4
cm); Sheet (paper): 21 1/2 × 14
1/8 in. (54.6 × 35.9 cm)
Purchased for the Society by
public subscription from Mrs.
John J. Audubon. New-York
Historical Society, 1863.17.185.

Upper Missouri River Region. Not only did the Blackfoot maintain their sovereignty and autonomy, they actively dictated terms to both the British and the Americans, thanks to their power in the fur trade.[110] Like Peale's imagined western landscapes, then, Audubon's portrait must be seen as a projection of colonialist desires, this time onto the person of Natoyist-Siksina'. Specifically, one might speculate that it was motivated by Audubon's discomfort with the multicultural environment at Fort Union. At the fort, identities were blurred, and the racial hierarchies to which the naturalist was accustomed were put into question.[111] Thus, Natoyist-Siksina's portrait became a means to restore and reassert what Audubon felt was the 'proper' racial and political order. Images once again demonstrate their power not to reveal worlds, but to *create* them, a power often wielded to advance the aims of empire.

Powerful in different ways is the watercolor study for plate 185, *Bachman's Warbler* of *The Birds of America*.[112] The original study is held in the collections of the New-York Historical Society and was represented in the exhibition by a fine art facsimile (Figure 10). The study is a testament to Audubon's long collaboration with the entire Bachman household, including the naturalist John Bachman, his sister-in-law and later wife Maria Martin, and enslaved members of their household, particularly a man called Thomas.[113] Audubon first met John Bachman in October 1831, when he stayed with the Bachmans in their Charleston home for a month-long visit. During this and Audubon's three subsequent residencies, in 1832, 1833–1834, and 1836–1837, the Bachman house was a major site of scientific collaboration, involving all members both enslaved and free. Even when Audubon was not in Charleston, Bachman shared specimens and information about southern bird species with Audubon.[114] During Audubon's first month-long visit, he also tutored Maria Martin in the art of botanical illustration.[115] Martin, a woman of considerable financial means who also enslaved individuals, lived with her sister and brother-in-law (Bachman) due to her sister's prolonged illness.[116] She was Audubon's only known female artist-collaborator, combining artistic skill and scientific precision to execute as many as 30 botanicals for volumes 2 and 4 of *Birds*.[117] Finally, after Audubon's visit a man called Thomas, enslaved by Bachman, began to taxidermy birds, suggesting that Audubon or his assistant Henry Ward provided Thomas with training during their initial stay.[118] The watercolor study for plate 185 brings together all of these collaborators. Audubon named the warbler for Bachman, who discovered it and sent specimens to Audubon. The depiction of *Gordonia pubescens* (the same plant that Bartram described as *Franklinia alatamaha*) is by Maria Martin, who in this instance executed the botanical watercolor first, with Audubon adding the birds in after.[119] Finally, Thomas's participation as taxidermist is more tenuous but highly plausible. Thomas used his skills as a taxidermist while working in the Bachman household, and Audubon based his depiction of the warbler on two preserved specimens sent to him by Bachman, as Audubon never actually saw this species in the wild.[120]

These varied contributions are sublimated under the overriding authorial name of Audubon, a process enhanced by the image's cohesiveness, thanks to Audubon's careful training of assistants to mimic his style. While artist's assistants during this period were commonly trained to suppress their own style in favor of their master's, Audubon also added to this dynamic, when he acknowledged their contributions in different ways and suppressed others.[121] The degree and type of credit Bachman, Martin, and Thomas are accorded speaks to the economic, racial, and gender hierarchies that structured the labor of natural history. In Bachman's case, Audubon prominently acknowledged him as the bird species' discoverer, naming the species after him. The eponym of *Bachman's Warbler* appears emblazoned on

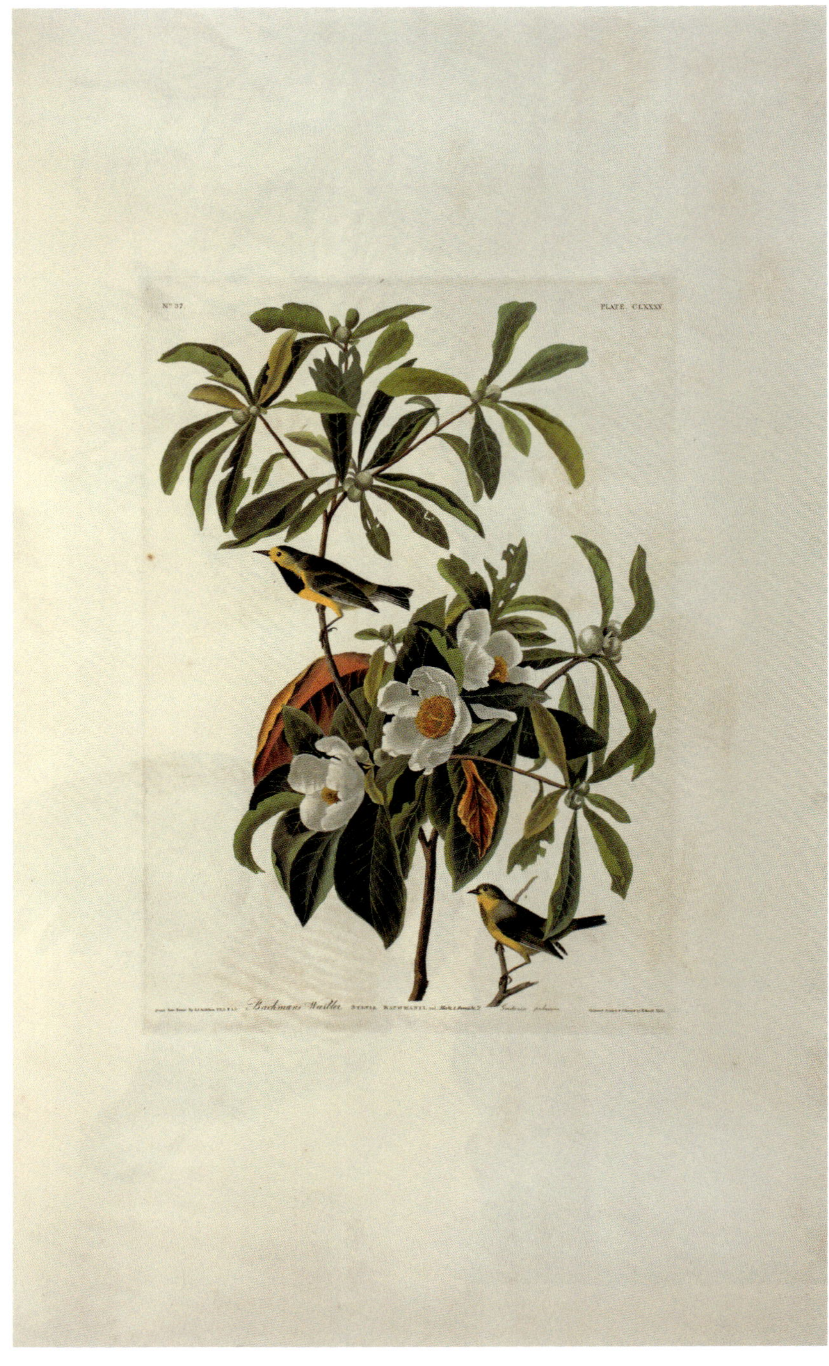

the published plate in volume two of *Birds of America* (1833) (Figure 11). Martin, on the other hand, is not acknowledged on the published plate, which is signed "Drawn from nature by J.J. Audubon" and "Engraved, Printed, and Coloured by R[obert] Havell," Audubon's printer. However, her botanical study is cited in the second volume of the *Ornithological Biographies* (1834), the textual descriptions for each species, which were published separately.[122] This separation of text and images allowed Audubon to skirt a British copyright law, but it also reflects Audubon's own emphasis on the images as the core of his project.[123] Audubon also only names Martin for nine of the approximately 30 plates to which she contributed.[124] Finally, Thomas gets no public mention at all, although his skills probably contributed not just to taxidermied specimens later represented by Audubon, but also to the care of a very large bird menagerie at the Bachman residence.[125] Moreover, Martin—a slaveholder herself—also no doubt relied on Thomas's labor during the hours she devoted to botanical illustration for Audubon.[126] The silence on Thomas's role underlines the many ways that Audubon exploited the system of plantation slavery to his own benefit.

The recent labor of scholars and curators has begun to correct the historical record around Audubon's collaborators. Yet the image itself offers the most potent refutation of this inaccurate history. If intertwined social hierarchies, communicated through text, shaped Bachman's very public celebration and Thomas's entirely unacknowledged contribution, the very structure of the image tells a different story. It is Martin's *Franklinia* that forms the luxuriant backdrop of the plate, and at the composition's very heart, within its rich green foliage, are two agile birds that may replicate Thomas's specimens. In short, the image has its own, formal hierarchy, and it is a powerful corrective to the historical distortions communicated through textual metadata.

Sketching Splendor reveals a period of natural history that is intellectually innovative and troubling all at once. It offers a truer picture of the legacies of William Bartram, Titian Ramsay Peale, and John James Audubon. Published texts, overlooked historical archives, as well as the labor of generations of scholars and curators all offer indispensable resources. But images, whether in their capacity to amplify discourses of inequality, to offer spaces for dialectical engagement, or even to correct the hierarchies they otherwise help to enact, also help us to illuminate this history.

NOTES

1   I give my sincere thanks to Amy Meyers, who read this essay and offered generous feedback. Alexis Wang and Lora Webb also read a draft and gave insightful advice. Any omissions or errors are my own.

2   The site is now known as Payne's Prairie. William Bartram, *The* Travels *of William Bartram: Naturalist's Edition,* ed. Francis Harper (New Haven: Yale University Press, 1958), 119–121.

3   Terms like 'ecology' and 'biodiversity' would not have been used by the naturalists in this exhibition, though they contributed to their emergence as intellectual concepts. 'Biodiversity' only came into use in the 1980s, 'ecology' was coined in 1866, for which see Robert P. McIntosh, *The Background of Ecology: concept and theory* (Cambridge: Cambridge University Press, 1985). The individuals treated in this exhibition would have described their subject as natural history, encompassing the study of the entire material world, including animals, plants, minerals, and more, which naturalists explored primarily by collecting and describing specimens. It was in this methodological sense that natural history was distinct from the contemporary practice of natural philosophy, which was concerned more strictly with causality and discovering natural laws. See Janet Browne, "Natural history," *The Oxford Companion to the History of Modern Science,* ed. John L. Heilbron (Oxford: Oxford University Press, 2003), 559–563.

4   Amy R.W. Meyers, "Sketches from the Wilderness: Changing Conceptions of Nature in American Natural History Illustration: 1680–1880" (PhD diss., Yale University, 1985), 113–151.

5   "On the first view of such an amazing display of the wisdom and power of the supreme author of nature, the mind for a moment seems suspended, and impressed with awe." Bartram, *Travels,* 120–121.

6   Thomas Hallock, "'On the Borders of a New World': Ecology, Frontier Plots, and Imperial Elegy in William Bartram's 'Travels'," *South Atlantic Review* 66 (Autumn 2001): 114.

7   Earlier in this same passage Bartram also notes both the immense scope and the "exuberantly fertile soil" of the savanna. Bartram, *Travels,* 119.

8   Paul L. Farber, *Discovering Birds: The Emergence of Ornithology as a Scientific Discipline, 1760–1850* (Baltimore and London: Johns Hopkins University Press, 1997), 7–26.

9   Farber, *Discovering Birds,* 23–26; Frank N. Egerton, *Roots of Ecology: Antiquity to Haeckel* (Berkeley and Los Angeles: University of California Press, Ltd., 2012), 84–86.

10 Meyers, "Sketches," 113–258; *ibid.,* "Observations of an American Woodsman: John James Audubon as Field Naturalist," in *John James Audubon: The Watercolors for* The Birds of America, eds. Annette Blaugrund and Theodore E. Stebbins, Jr. (New York: New-York Historical Society, 1993), 43–54; see also Sue Ann Prince, "Introduction," in *Stuffing Birds, Pressing Plants, Shaping Knowledge: Natural History in North America, 1730–1860,* ed. Sue Ann Prince (Philadelphia, PA: American Philosophical Society, 2003), 2–4; Susan Scott Parrish, *American Curiosity: Cultures of Natural History in the Colonial British Atlantic World* (Chapel Hill, NC: Published for the Omohundro Institute of Early American History and Culture, Williamsburg, Virginia, by the University of North Carolina Press, 2006), 103–135.

11 Ann Shelby Blum, *Picturing Nature: American Nineteenth-century Zoological Illustration* (Princeton: Princeton University Press, 1993), 20–118; Kenneth Haltman, *Looking Close and Seeing Far: Samuel Seymour, Titian Ramsay Peale, and the Art of the Long Expedition, 1818–1823* (University Park, PA: Pennsylvania State University Press, 2008), 1–30.

12 Mary Louise Pratt, *Imperial Eyes: Travel Writing and Transculturation,* 2nd ed. (London and New York: Routledge, 2008), 24–66; Richard Drayton, *Nature's Government: Science, Imperial Britain, and the 'Improvement' of the World* (New Haven: Yale University Press, 2000); Christopher M. Parsons, *A Not-So-New World: Empire and Environment in French Colonial North America* (Philadelphia: University of Pennsylvania Press, 2018); on natural knowledge and plantation slavery, Cameron B. Strang, *Frontiers of Science: Imperialism and Natural Knowledge in the Gulf South Borderlands, 1500–1850* (Chapel Hill: University of North Carolina Press, 2018), 245–286.

13 For discussions of Audubon and Bartram, see below. See also Gwendolyn Dubois Shaw, "'Moses Williams, Cutter of Profiles': Silhouettes and African American Identity in the Early Republic," *Proceedings of the American Philosophical Society* 149 (March 2005): 22–39.

14 J. Drew Lanham, "What Do We Do About John James Audubon?," *Audubon Magazine* (Spring 2021), https://www.audubon.org/magazine/spring-2021/what-do-we-do-about-john-james-audubon; Gregory Nobles, "The Myth of John James Audubon," Audubon, July 31, 2020, https://www.audubon.org/news/the-myth-john-james-audubon.

15 Parrish, *American Curiosity*; Janine Yorimoto Boldt, *Dr. Franklin: Citizen Scientist* (Philadelphia, PA: American Philosophical Society Press, 2020), 10–40.

16 This is thanks in no small part to collaboration with generous outside researchers and institutions. In particular, I benefitted enormously from conversations with E. Bennett Jones, as well as their dissertation, currently under revision for publication, E. Bennett Jones, "'The Indians Say': Settler Colonialism and the Scientific Study of North America, 1722 to 1848" (PhD diss., Northwestern University, 2021). I would also like to thank staff members of the University of Michigan and University of Michigan Library, Marieka Kaye, Juli McLoone, Caitlin Pollock, and Jason Young, who shared working documents related to their own research into the collaborators of John James Audubon that directly informed the exhibition's content. See note 98.

17 Meyers, "Sketches," 2–3; Blum, *Picturing Nature*, 88, 112; Sachiko Kusukawa, "Leonhart Fuchs on the Importance of Pictures," *Journal of the History of Ideas* 58 (July 1997): 403–427; Michael Gaudio, "Surface and Depth: The Art of Early American Natural History," in *Stuffing Birds, Pressing Plants, Shaping Knowledge: Natural History in North America, 1730–1860*, ed. Sue Ann Prince (Philadelphia: American Philosophical Society, 2003), 55–74; Ellery Foutch, "Capturing Nature: American Artists' Pursuit of Natural History," in *Flora/Fauna: The Naturalist Impulse in American Art*, ed. Jennifer Stettler Parsons (Old Lyme, CT: Florence Griswold Museum, 2017), 75–96.

18 Edouard Kopp, Elizabeth M. Rudy, and Kristel Smentek, "Introduction," in *Dare to Know: Prints and Drawings in the Age of Enlightenment*, eds. Edouard Kopp, Elizabeth M. Rudy, and Kristel Smentek, exhibition catalog (Cambridge, MA: Harvard Art Museums, 2022), 2–3.

19 Thomas Hallock, "Signing Nature, Memorializing Plantations: Public Memory along the William Bartram Trail," in *The Attention of a Traveller: Essays on William Bartram's* Travels *and Legacy*, ed. Kathryn H. Braund (Tuscaloosa: The University of Alabama Press, 2022), 198; Pratt, *Imperial Eyes*, 37–65.

20 The garden was also a formative influence on William's ecological world view. See Therese O'Malley, "Cultivated Lives, Cultivated Spaces: The Scientific Garden in Philadelphia, 1740–1840," in *Knowing Nature: Art and Science in Philadelphia, 1740–1840*, ed. Amy R.W. Meyers with Lisa L. Ford (New Haven and London: Yale University Press, 2011), 36–59; and Joel T. Fry, "America's 'Ancient Garden': The Bartram Botanica Garden, 1728–1850," in *ibid.*, 60–95.

21 Raymond Phineas Stearns, *Science in the British Colonies of America* (Chicago: University of Illinois Press, 1970), 575–592.

22 Nancy Everill Hoffmann, "The Construction of William Bartram's Narrative Natural History: A Genetic Text of the Draft Manuscript for *Travels through North & South Carolina, Georgia, East & West Florida*" (PhD diss., University of Pennsylvania, 1996), 3; Judith Magee, *The Art and Science of William Bartram* (University Park, PA: Pennsylvania State University Press, 2007), 8; Christoph Irmscher, *The Poetics of Natural History*, with Rosamond Purcell, 2nd edition (New Brunswick, NJ: Rutgers University Press, 2019), 17–63; Peter C. Messer, "The Nature of William Bartram's *Travels*," in *Atlantic Environments and the American South*, eds. Thomas Blake Earle and D. Andrew Johnson (Athens, GA: University of Georgia Press, 2020), 195–214.

23 Kerry S. Walters, "The 'Peaceable Disposition' of Animals: William Bartram on the Moral Sensibility of Brute Creation," *Pennsylvania History* 56 (July 1989): 157–176; Hallock, "On the Borders," 109–133; *ibid.*, "Signing Nature," 196–215; Christopher Iannini, "The Vertigo of Circum-Caribbean Empire: William Bartram's Florida," *The Mississippi Quarterly* 57 (Winter 2003–4): 147–156; *ibid.*, *Fatal Revolutions: Natural History, West Indian Slavery, And the Routes of American Literature* (Chapel Hill: Published for the Omohundro Institute of Early American History and Culture, Williamsburg, Virginia, by the University of North Carolina Press, 2012), 177–217; Magee, *The Art and Science*, 8, 97–122, 125–132, 134–140, 149, 212–214; Richard W. Judd, *The Untilled Garden: Natural History and the Spirit of Conservation in America, 1740–1840* (New York: Cambridge University Press, 2009), 183–214; Messer, "The Nature," 195–214.

24 Hallock, "On the Borders," 111–112; Hoffman, "The Construction," 1–48. As Iannini notes, arguably its genesis began even earlier, in the 1765 expedition William undertook with his father, John Bartram, recently appointed Botanist to the King. William supported British efforts to secure control over its new territory in East Florida through the establishment of a 500-acre rice plantation, Iannini, "The Vertigo," 150.

25 Hallock, "On the Borders," 114.

26 Magee, *The Art and Science,* 98; Iannini, *Fatal Revolutions,* 206–209; Messer, "The Nature," 195–214.

27 Meyers, "Sketches," 132–151; Gaudio, "Surface and Depth," 70–71; Elizabeth Athens, "Figuring a World: William Bartram's Natural History" (PhD diss., Yale University, 2015), 69–73. As a number of scholars have noted, there is a tight relationship between the image and Bartram's two main descriptions of the Alachua Savanna in the *Travels,* Bartram, *Travels,* 119–132, 158. On these text-image relationships, for example, Iannini, *Fatal Revolutions,* 207.

28 For the analysis of these spatial dynamics in Bartram's image, I draw on Meyers, "Sketches," 132–152.

29 I am paraphrasing Iannini, "The Vertigo," 152; see also Iannini, *Fatal Revolutions,* 207–209.

30 Meyers, "Sketches," 145–151; Magee, *The Art and Science,* 134–140.

31 Joyce E. Chaplin, "Nature and Nation," in *Stuffing Birds, Pressing Plants, Shaping Knowledge: Natural History in North America, 1730–1860,* ed. Sue Ann Prince (Philadelphia, PA: American Philosophical Society, 2003), 77.

32 Compare with Bartram's own map, serving as the frontispiece for the original 1791 edition. On its map-like qualities, Gaudio, "Surface and Depth," 70–71 and Athens, "Figuring," 69–73.

33 Interestingly, in the right half of the image Bartram has also correctly placed the village of Cuscowilla to the southeast of the savanna, whereas in the 1775 map for John Fothergill he has incorrectly placed it to the northeast. Athens, "Figuring," 73.

34 Joseph Ewan, *William Bartram: Botanical and Zoological Drawings, 1756–1788* (Philadelphia: The American Philosophical Society, 1968), 83. The chronological relationship between these two drawings is not clear.

35 I found the textual analysis in Messer, "The Nature" especially helpful, though I maintain a stronger sense of Bartram's ambivalence as opposed to his resolution of tensions.

36 In the passages immediately following Bartram's famous exclamations, he turns back to the more conventional language of natural history, describing animal and plant life (Bartram, *Travels,* 122, 123, 127, 128) and assessing the savanna's terrain (*ibid.,* 125, 126, 128, 132). Descriptions of the terrain include commentary on the alternately fertile ground and boggy morasses clearly depicted in the APS image. This is all part of what was clearly a coordinated survey of the site undertaken by Bartram and the group of traders with whom he travels, "Early next morning we continued our tour; one division of our company directing their course across the plains to the North: my old companion, with myself in company, continued our former route, coasting the savanna W. and N.W., and by agreement we were all to meet again at night, at the E. end of the savanna" (*ibid.,* 126). Moreover, on Bartram's second trip to the savanna, in June of 1774, he very clearly anticipates its appropriateness for a large European-style settlement, Hallock, "On the Borders," 130; Gregory Waselkov and Kathryn E. Holland Braund, *William Bartram on the Southeastern Indians* (Lincoln, NE: University of Nebraska Press, 1995), 202.

37 Gaudio, "Surface and Depth," 71 remarks on this dialectic.

38 Waselkov and Braund, *William Bartram,* 202–4; Magee, *The Art and Science,* 125–32; Kathryn E. Holland Braund, "Native Americans in William Bartram's 'Hints & Observations'," in *William Bartram, the Search for Nature's Design: selected art, letters, and unpublished writings,* eds. Thomas Hallock and Nancy E. Hoffmann (Athens, GA: The University of Georgia Press, 2010), 359–371.

39 Clearly, which power was doing the annexing shifted between Bartram's 1774 visit to the savanna and the 1791 publication of *Travels.*

40 "….the great and beautiful Alachua Savanna, whose exuberant green meadows, with the fertile hills which immediately encircle it, would if peopled and cultivated after the manner of the civilized countries of Europe, without crouding [sic] or incommoding families, at a moderate estimation,

accommodate in the happiest manner, above one hundred thousand human inhabitants, besides millions of domestic animals; and I make no doubt this place will at some future day be one of the most populous and delightful seats on earth." Bartram, *Travels*, 158. Analyzed and quoted in Hallock, "On the Borders," 129; Waselkov and Braund, *William Bartram*, 202, 243. For another similar passage, Iannini, "The Vertigo," 201.

41 Waselkov and Braund, *William Bartram*, 202, 243.

42 The document is "Some Hints & Observations, concerning the civilization, of the Indians, or Aborigines of America," see analysis and edition in Braund, "Hints and Observations," 359–71.

43 *Ibid.*, 364, 371.

44 For the text and a brief analysis, Kerry Walters, "All Equally Dear to God: William Bartram's Anti-slavery Manuscripts," in *William Bartram: The Search for Nature's Design*, eds. Thomas Hallock and Nancy E. Hoffmann (Athens, GA: The University of Georgia Press, 2010), 372–380. Walters dates the treatise after 1787 and around 1790, based on his belief Bartram refers to the Constitution. However, Joel Fry believed the language referenced the Articles of Confederation and thus could date to the mid-1780s. See Joel T. Fry, "Slavery and Freedom at Bartram's Garden," (lecture, Investigating Mid-Atlantic Plantations: Slavery, Economies, and Space Conference, Philadelphia, Pennsylvania, 17–19 October 2019) https://www.bartramsgarden.org/wp-content/uploads/J.Fry-slavery-freedom-at-Bartrams-Garden-MCEAS-2019-4.pdf. Content by Joel T. Fry (1957–2023), courtesy of the John Bartram Association, Bartram's Garden, Philadelphia. All rights reserved. Fry's research into the Bartrams' reliance on enslaved labor was built on the work of Sharece Blakney, *Stories We Know: Recording the Black History of Bartram's Garden and Southwest Philadelphia*, ed. Aislinn Pentecost-Farren (Philadelphia: John Bartram Association/Mural Arts Philadelphia, 2017).

45 I drew on the following resources when reconstructing Bartram's complicated and shifting views on slavery: Magee, *The Art and Science*, 71–75; Fry, "Slavery and Freedom"; Hallock, "Signing Nature"; Joel T. Fry and Amy Meyers, "William Bartram (1739–1823), *Catalogue of American Trees, Shrubs and Herbaceous Plants, most of which are now growing, and produce ripe Seed in John Bartram's Garden, near Philadelphia*," in *Two Hundred Years: The Historical Society of Pennsylvania, 1824–2024*, ed. David R. Brigham (Philadelphia: The Historical Society of Pennsylvania, 2023), 92–94.

46 Hallock, "Signing Nature," 200. For the letter describing the sale, negotiated by William's brother and brother-in-law, Hallock and Hoffman, *Search for Nature*, 97. The bill of sale is in the American Philosophical Society Collections, Collinson-Bartram Papers, 1732–1773, Mss.B.C692.1, "Bill of Sale for a Slave."

47 Iannini, *Fatal Revolutions*, 196.

48 Magee, *The Art and Science*, 75.

49 See Fry and Meyers, "*Catalogue of American Trees*," 92–94.

50 Though the historical documentation is not clear, there was probably at least one enslaved laborer at the garden, who was later manumitted. Fry, "Slavery and Freedom."

51 Fry and Meyers, "Catalogue of American Trees," 94.

52 Magee, *The Art and Science*, 184–187.

53 For Titian's participation in an 1818 expedition organized by the Academy of Natural Sciences and intended to retrace Bartram's Florida journey, Charlotte M. Porter, "Following Bartram's 'Track': Titian Ramsay Peale's Florida Journey," *The Florida Historical Quarterly*, 61 (April 1983): 431–444.

54 Magee, *The Art and Science*, 179–180.

55 Peale's Philadelphia Museum became the unofficial repository for specimens acquired during govern-ment-sponsored expeditions, beginning with the Lewis & Clark Expedition. See Kenneth Haltman, "The Pictorial Legacy of Lewis and Clark," in *Knowing Nature: Art and Science in Philadelphia, 1740–1840*, ed. Amy R.W. Meyers with Lisa L. Ford (New Haven: Yale University Press, 2011), 333. On the museum's locations and its organization, David R. Brigham, *Public Culture in the Early Republic: Peale's Museum and Its Audience* (Washington, DC: Smithsonian Institution Press, 1995), 13–7, 44–47, 58–60, 64–67.

56 Specimens from the Long Expedition as well as drawings were also to be placed on deposit at Peale's Philadelphia Museum, Charlotte M. Porter, "The Lifework of Titian Ramsay Peale," *Proceedings of the American Philosophical Society* 129 (1985): 307; Haltman, *Looking Close*, 158. For overviews of Titian's work see also Robert Cushman Murphy, "The Sketches of Titian Ramsay Peale (1799–1885)," *Proceedings of the American Philosophical Society* 101 (Dec. 19, 1957): 523–31; Jessie Poesch, *Titian Ramsay Peale, 1799–1885, And His Journals of the Wilkes Expedition* (Philadelphia: American Philosophical Society, 1961).

57 Poesch, *Titian Ramsay Peale*, 23–55.

58 Haltman, "The Pictorial Legacy," 330–355.

59 This included Thomas Say, as zoologist, entomologist, ornithologist, and ethnographer; Augustus Jessup as geologist; William Baldwin as botanist; Edwin James, botanist, who replaced both Jessup and Baldwin in the spring of 1820; and Samuel Seymour, landscape painter. Stephen Long was to lead the expedition and also serve as its cartographer, with William Swift, assistant cartographer. Titian Ramsay Peale's official title was assistant naturalist. Edwin James, *Account of an Expedition from Pittsburgh to the Rocky Mountains...* (Philadelphia: H.C. Carey and I. Lea, 1822), vol 1: 425; Haltman, "The Pictorial Legacy," 342–343.

60 James, *Account*, vol 1: 3–4.

61 See Amy Meyers, whose interpretations I follow here, Meyers "Sketches," 194–258 and *ibid.*, "Imposing Order on the Wilderness: Natural History Illustration and Landscape Portrayal," in *Views and Visions: American Landscape before 1830*, ed. Edward J. Nygren (Washington, D.C.: The Corcoran Gallery of Art, 1986), 104–131. See also the trenchant analysis of this image by Haltman, *Looking Close*, 7–9.

62 Haltman, *Looking Close*, 7–9.

63 Judd, *The Untilled Garden*, 183–214; Amy R.W. Meyers, "From nature and memory: William Bartram's drawings of North American flora and fauna," in *Knowing Nature: Art and Science in Philadelphia, 1740–1840*, ed. Amy R.W. Meyers with Lisa L. Ford (New Haven: Yale University Press, 2011), 128–59.

64 Meyers "Sketches," 194–258 and *ibid.*, "Imposing Order," 128–131.

65 Meyers, "Sketches," 244–245.

66 Meyers, "Sketches," 258.

67 Anne F. Hyde, "Cultural Filters: The Significance of Perception in the History of the American West," *The Western Historical Quarterly* 24 (Aug. 1993): 351–374, esp. 355; Meyers, "Imposing Order," 123–130; on the transformation of non-European landscapes according to European taxonomic conventions, Pratt, *Imperial Eyes*, 24–36. For Peale's artistic process as quest for individual control over nature, Haltman, *Looking Close*, 109–132.

68 Hyde, "Cultural Filters," 355; Meyers "Sketches," 247. Only the diaries from the first three months are extant, see, A.O. Weese, ed. "The Journal of Titian Ramsay Peale," *Missouri Historical Review* 41 (January 1947): 147–163, (April 1947): 266–284.

69 Meyers, "Sketches," 255–258; Richard H. Dillon, "Stephen Long's Great American Desert," *Proceedings of the American Philosophical Society* 111 (Apr. 14, 1967): 93–108.

70 The *Account* was published in both Philadelphia and London. The Philadelphia edition consisted of two octavo volumes of text and one quarto atlas. The entire Philadelphia edition of the *Account* was published in 1822, despite the 1823 date on its frontispiece. The London edition, published in 1823, consisted of three octavo volumes of text with plates interleaved. There are differences across the text and images in these two editions, see Neal Woodman, "History and dating of the publication of the Philadelphia (1822) and London (1823) editions of Edwin James's *Account of an expedition from Pittsburgh to the Rocky Mountains*," *Archives of Natural History* 37 (2010): 28–38.

71 The *Dusky Wolf* records a museum display, a composite that made use of a partially decomposed deer specimen, Haltman, *Looking Close*, 165. See also Kenneth Haltman, "Private Impressions and Public Views: Titian Ramsay Peale's Sketchbooks from the Long Expedition, 1819–1820," *Yale University Art Gallery Bulletin* (Spring 1989): 38–53.

72 On obscurity—intellectual and visual—in Seymour's watercolors, Haltman, *Looking Close*, 30–54.

73 Meyers, "Sketches," 227.

74 Parrish, *Curiosity*, 77–102. Peale was directly involved in pseudo-scientific studies of human difference. While on the Long Expedition, he looted and stole human remains from two Native American
burial grounds, lending one skull to Samuel G. Morton (APS, 1828), the Philadelphia physician who
assembled a massive collection of human remains as part of his white supremacist effort to 'prove' the
superiority of Europeans. James, *Account*, vol 1: 62–66, 458; Samuel G. Morton, *Crania Americana; or,
A Comparative View of the Skulls of Various Aboriginal Nations of North and South America: To which is
Prefixed An Essay on the Varieties of the Human Species* (Philadelphia: J. Dobson, 1839), 196. Morton says
that Peale lent him the skull. The Peale Museum also displayed wax figures of different human races
in a section devoted to archaeological and ethnographic materials that relegated nonwhite cultures
to an ancient past. Brigham, *Public Culture*, 125–144, Charles Coleman Sellers, *Mr. Peale's Museum:
Charles Willson Peale and the First Popular Museum of Natural Science and Art* (New York: W.W. Norton
& Company, Inc., 1980), 162. Peale also displayed physical human remains obtained as war booty by
American soldiers during the Revolutionary and Northwest Indian Wars, see Ellen Fernandez-Sacco,
"Framing 'The Indian': The Visual Culture of Conquest in the Museums of Pierre Eugene Du Simitiere
and Charles Willson Peale, 1779–96," *Social Identities* 8 (2002): 571–618.

75 James, *Account*, vol 2: 125; for a different reading of Peale and Seymour's depictions of Native American
communities, Haltman, *Looking Close*, 180–181.

76 The Philadelphia edition atlas included two maps and two geological elevations by Long and Swift,
six images by landscape painter Samuel Seymour, and two images by Peale. The two images by Peale
are both ethnographic, a Kaskaia tipi and a painted bison robe depicting a battle scene. Peale's two
images were also published in the London edition. Peale's natural history illustrations were available
on deposit at the Peale Museum, as well as informing numerous museum displays, which Peale
constructed using specimens he had gathered. Some were also published later in different contexts.
Poesch, *Titian Ramsay Peale*, 37, 43; Porter, "The Lifework," 307; Haltman, *Looking Close*, 158. For the
Philadelphia versus London editions, see note 69 above.

77 Dillon, "Stephen Long," 93.

78 James, *Account*, vol 1: 426–427, 435–443.

79 James refers to these different bands as, respectively, the "Grand" Pawnees, the "Republican" Pawnees,
and the "Loup" Pawnees. I follow the terminology used today by the Pawnee Nation.

80 Poesch, *Titian Ramsay Peale*, 33.

81 Jones, "The Indians Say," 11.

82 Dillon, "Stephen Long," 104.

83 The identity of Audubon's mother is not known with certainty. She may have been a chambermaid
who worked in Audubon Sr.'s household called Jeanne Rabine, who herself was likely from Nantes,
France. However, Audubon Sr. had children with multiple women during his time in Saint-
Domingue, including a mixed-raced woman called Catherine 'Sanitte' Bouffard. For Audubon's basic
biography: Roberta J.M. Olson, "A Biographical Sketch of an American Icon," in *Audubon's Aviary:
The Original Watercolors for* The Birds of America (New York: New-York Historical Society, 2012),
16–39. For divergent proposals regarding the racial identity of Audubon's mother: Gregory Nobles,
John James Audubon: The Nature of the American Woodsman (Philadelphia: University of Pennsylvania
Press, 2017), 8–26; Roberta J.M. Olson, "Hiding in Plain Sight: New Evidence About the Birth,
Identity, and Strategic Pseudonyms of John James Audubon," *Bulletin of the Museum of Comparative
Zoology* 163 (Sept. 2021): 129–150.

84 Meyers, "Observations," 46.

85 Annette Blaugrund and Theodore E. Stebbins, Jr., eds., *John James Audubon: The Watercolors for* The
Birds of America (New York: New-York Historical Society, 1993); Blum, *Picturing Nature*, 88–118;
Linda Dugan Partridge, "By the Book: Audubon and the Tradition of Ornithological Illustration," in
Art and Science in America: Issues of Representation, ed. Amy R.W. Meyers (San Marino, CA: Henry E.
Huntington Library and Art Gallery, 1998), 97–130; Irmscher, *The Poetics*, 199–245; Diana Donald, "The
'Struggle for Existence' in Nature and Human Society," in *Endless Forms: Charles Darwin, Natural
Science and the Visual Arts,* eds. Diana Donald and Jane Munro (New Haven, London: Yale University

Press, 2009), 81–99; Roberta J.M. Olson, "Audubon's Innovations and the Traditions of Ornithological Illustration: Some Things Old, Some Things Borrowed, But Most Things New," in *Audubon's Aviary: The Original Watercolors for* The Birds of America, ed. Roberta J.M. Olson (New York: New-York Historical Society, 2012), 40–106; Jennifer L. Roberts, *Transporting Visions: The Movement of Images in Early America* (Berkeley, Los Angeles, London: University of California Press, 2014), 76–110; Alan C. Braddock, "The Order of Things," in *Nature's Nation: American Art and Environment,* eds. Karl Kusserow and Alan C. Braddock (Princeton: Princeton University Art Museum, 2018), 43–69. For evidence that Audubon also fabricated scientific evidence to advance his career, Matthew R. Halley, "Audubon's Bird of Washington: Unravelling the Fraud that launched *The Birds of America*," *Bulletin of the British Ornithologists' Club* 140 (2020): 110–141.

86 This is due to their association with the Audubon Society, a conservation founded after Audubon's death by George Grinnell in 1886 and then reestablished in 1895 by Harriet Hemenway and Mina Hall. Grinnell had been tutored by Lucy Audubon, Nobles, *John James Audubon,* 264–268. For Audubon's conflicted status as a proto-conservationist, see Daniel Patterson, "Audubon's Conservation Ethic Reconsidered," in *The Missouri River Journals of John James Audubon* (Lincoln and London: University of Nebraska Press, 2016), 211–304, and p. 216 for one particularly alarming quotation, in a letter Audubon wrote from Florida on Dec 31, 1831, "The birds, generally speaking, appeared wild and few— you must be aware that I call birds few, when I shoot less than one hundred per day."

87 Lanham, "What Do We Do;" Nobles, "The Myth." For a recent example from the museum world, see Marieka Kaye, Juli McLoone, Caitlin Pollock, and Jason Young, "Revisioning 'The Birds of America' at the University of Michigan Library," University of Michigan Library, November 2022, https://www.lib.umich.edu/static/856bcb7b218f03a282821802a6e2f3f6/The-Birds-of-America_0.pdf.

88 Nobles, *John James Audubon,* 200–208. He privately disapproved of the British government's decision to abolish slavery in the British West Indies in 1834, American Philosophical Library Collections, John James Audubon Papers, Mss.B.Au25, John James Audubon to Lucy B. Audubon, September 5–11, 1834, "…in giving Freedom to the Slaves…the British Government…acted imprudently and too precipitously."

89 Audubon sent Morton the remains of five Mexican, Hispano-Indian, and Native American individuals stolen during his 1837 Texas expedition and five more from his 1843 Missouri Expedition. Matthew R. Halley, "The literal skeletons in the closet of American ornithology," June 6, 2020, https://matthewhalley.wordpress.com/2020/06/16/the-literal-skeletons-in-the-closet-of-american-ornithology/; Ann Fabian, "We left all on the ground but the head: JJ Audubon's Human Skulls," *Commonplace* (2021) https://commonplace.online/article/audubons-human-skulls/; Samuel G. Morton, *Catalogue of the Skulls of Man and the Inferior Animals in the Collection of Samuel George Morton* (Philadelphia: Turner and Fisher, 1840), nos. 555–558, 690; Samuel G. Morton, *Catalogue of the Skulls of Man and the Inferior Animals in the Collection of Samuel George Morton* (Philadelphia: Turner and Fisher, 1849), nos. 555–558, 690, nos. 1227–1231. Archival documents in APS collections appear to record approximately five more individuals whose remains Audubon sent to Morton. It is difficult to provide a definitive number. See American Philosophical Society Collections, Samuel George Morton Papers, Mss.B.M843, John Bachman to Samuel G. Morton, March 17, 1837, "two Indian chief skulls…one…owned by Audubon;" John James Audubon to Samuel G. Morton, July 16, 1837, "The Yazoo has not reached its haven….I have left all your wishes for skulls…into the hands of my….friend Ed Harris….;" John James Audubon to Samuel G. Morton, August 28, 1837, "I am….awaiting to hear….if my friend Ed Harris forwarded you the skulls…;" John James Audubon to Samuel G. Morton, June 25, 1838, "The two Indian skulls which I have for you…."

90 The imperial folio edition of *Quadrupeds*, comprising just plates as in the *Birds*, was issued between 1845–1848; the accompanying text between 1846–1854; and an octavo edition, combining both images and text, between 1849–1854. His sons published further editions. Charles T. Butler, ed., *Audubon's Last Wilderness Journey: The Viviparous Quadrupeds of North America* (Auburn, AL: Julie Collins Smith Museum of Fine Art, Auburn University, 2018), 22.

91 Blum, *Picturing Nature,* 118; John R. Knott, *Imagining Wild America* (Ann Arbor: University of Michigan Press, 2002), 17–48; Nobles, *John James Audubon,* 5, 173, 199–200, 200–208; Theodore E. Stebbins, Jr., "Audubon's Drawings of American Birds, 1805–38," in *John James Audubon: the Watercolors for* The Birds of America, eds. Annette Blaugrund and Theodore E. Stebbins, Jr. (New York: New-York Historical Society, 1993), 6–7.

92 Lanham, "What Do We Do," makes this point. See also Nobles, *John James Audubon,* 8–26; Olson, "Hiding in Plain Sight," 129–150. For Audubon's discomfort with his French Creole identity and how it may have motivated some of his work, Iannini, *Fatal Revolutions,* 253–279.

93 Annette Blaugrund, "The Artist as Entrepreneur," in *John James Audubon: The Watercolors for* The Birds of America, eds. Annette Blaugrund and Theodore E. Stebbins, Jr. (New York: New-York Historical Society, 1993), 35–37, 40–41; Olson, "A Biographical Sketch," 25–27, 29–30, 32–34; Nobles, *John James Audubon,* 109–115, 161–162. Audubon relied particularly hard on collaborators during the *Quadrupeds,* only providing half the paintings and unfinished text from his journals, Butler, *Audubon's Last Journey,* 14.

94 Nobles, *John James Audubon,* 49–50, 211–213, 219–224; Debra J. Lindsay, *Maria Martin's World: Art & Science, Faith & Family in Audubon's America* (Tuscaloosa, Alabama: The University of Alabama Press, 2018), 9–10, 16, 46–112; Jones, "The Indians Say," 125–163. Juli McLoone's research for the University of Michigan Library outlines some of Audubon's Native American, African, women, and other collaborators, including Owen Mackenzie, Maria Martin, and Thomas. See Juli McLoone, "Making 'The Birds of America'," University of Michigan Library, November 2022, https://www.lib.umich.edu/static/856bcb7b218f03a282821802a6e2f3f6/The-Birds-of-America_0.pdf.

95 I would also like to thank our two Museum Interns, Lauren McCouch and Spencer Auerbach, for their research into Audubon's collaborators.

96 My thanks to Jennifer Spence, Curator, Kentucky State Parks, for generously allowing me to examine the portraits and relevant curatorial files. The portrait is inscribed on the back, in a 19th-century script, "[To] John Culbertson from his cousin Natawista. 1857." Culbertson and Natoyist-Siksina' gave the portraits to Culbertson's cousin, John Moodey Culbertson, in 1857 and they remained in the Culbertson family until 1984. The portrait and its pendant have sometimes been attributed to John Mix Stanley, who depicted Natoyist-Siksina' in the 1850s. I believe that both portraits are properly ascribed to Audubon based on stylistic comparison to Audubon's portraits of his sons, *Victor Gifford Audubon* and *John Woodhouse Audubon* of c. 1823, both also in the collection of John James Audubon State Park. All four portraits exhibit an uncertainty in the oil medium that is consistent with Audubon's preference for watercolor, pastel, and mixed media. Stanley was a professional painter who worked frequently in oil. All the portraits under discussion also share a characteristic tendency to enlarge the head and eyes in relation to the body that is also visible in Audubon's crayon and charcoal portraits.

97 Alexander Culbertson had married a Piikani woman in 1833, whose name is unknown and whom he probably divorced when he married Natoyist-Siksina'. Ryan Hall, *Beneath the Backbone of the World: Blackfoot People and the North American Borderlands, 1720–1877* (Chapel Hill: The University of North Carolina Press, 2020), 111. For basic biography, Hugh A. Dempsey, "Natawista (Natoyist-siksina, Natawista Iksana, known as Medicine Snake Woman)," *Dictionary of Canadian Biography,* vol. XII (1891–1900), http://www.biographi.ca/en/bio/natawista_12E.html.

98 This was Audubon's only expedition for *Quadrupeds.* The group comprised Audubon's friend Edward Harris, John Graham Bell as taxidermist, artist Isaac Sprague, and Lewis Squires as secretary and aide. They arrived on June 12 and stayed two months. At their departure they were accompanied by Alexander Culbertson, Natoyist-Siksina', and their infant, as Culbertson had been transferred to another post on the Platte River. See John James Audubon, "Missouri River Journals," in *John James Audubon: Writings and drawings,* ed. Christoph Irmscher (New York: Literary Classics of the United States, 1999), 629, 731–732, 734. Secondary sources, Mary Durant, *On the Road with John James Audubon* (New York: Dodd, Mead, 1980), 572–574; Michael Harwood, "Mr. Audubon's Last Hurrah," *Audubon* 87 (November 1985): 98, 101–110; Barton H. Barbour, *Fort Union and the Upper Missouri Fur Trade* (Norman: University of Oklahoma Press, 2001), 82–87; Butler, *Audubon's Last Journey,* 14; Hall, *Beneath the Backbone,* 112.

99 Audubon states that Natoyist-Siksina' gave him six mallards she had caught herself and noted that she caught a Golden Eagle in order to use its feathers in a parfléche. For these and other relevant passages, Audubon, "Missouri River," 673, 697, 734. For an argument that, during the winter of 1810–1811, Audubon benefitted from a multicultural environment facilitated by marriage alliances between French fur traders and elite Shawnee women, Jones, "Indians Say," 135–141. Jones argues, however, that by 1843 Audubon's hardening racism made him less receptive to Indigenous knowledge, Jones, "Indians Say," 157, 162.

100 For example, when Culbertson and Natoyist-Siksina' were reassigned in summer 1843 to a new post on the North Platte River, Fort Laramie, relations with the Blackfoot Confederacy on the upper Missouri immediately declined, and ultimately they were brought back to reestablish relations in fall 1845. She also played an important role in the negotiations between the United States Government and the Blackfoot Confederacy that ultimately led to an 1855 treaty. Hall, *Beneath the Backbone,* 110–113, 128–139.

101 In summer 1843 Audubon also describes Natoyist-Siksina' meeting with a group of mixed-race Native American visitors to Fort Union in her private quarters, Audubon, "Missouri River," 705.

102 Harwood, "Last Hurrah," 101–110.

103 McKenzie appears regularly in Audubon's account from their first days at Fort Union, for his first appearance, Audubon, "Missouri River," 639. Jones, "Indians Say," 157. Owen Mackenzie was the son of Kenneth McKenzie, a fur trader with the American Fur Company who negotiated its expansion into the Upper Missouri River region with the Blackfoot, and an unidentified Native American woman. Hiram M. Chittenden, *The American Fur Trade of the Far West* (New York: Francis P. Harper, 1902), 1: 384–387.

104 Audubon, "Missouri River," 669, 674.

105 Audubon, "Missouri River," 677.

106 Eva Lajer-Burcharth, "Human," in *Dare to Know: Prints and Drawings in the Age of Enlightenment,* eds. Edouard Kopp, Elizabeth M. Rudy, and Kristel Smentek, exhibition catalog (Cambridge, MA: Harvard Art Museums, 2022), 90–91.

107 For example, Audubon projects these perceived tensions onto Natoyist-Siksina's choice of food, "I lost the head of my first bull because I forgot to tell Mrs. Culbertson that I wished to save it, and the princess had its skull broken open to enjoy its brains. Handsome, and really courteous and refined in many ways, I cannot reconcile to myself the fact that she partakes of raw animal food with such evident relish." Audubon, "Missouri River," 696.

108 The backgrounds may be the work of Audubon's assistant Isaac Sprague, as suggested to me by Jennifer Spence (personal communication).

109 For these ideas in John Mix Stanley's depictions of Natoyist-Siksina', Brian W. Dippie, "The Moving Finger Writes: Western Art and the Dynamics of Change," in *Discovered Lands Invented Pasts,* ed. Jules David Prown et al. (New Haven: Yale University Press, 1992), 101–102.

110 Hall, *Beneath the Backbone,* 89–115, esp. p. 114.

111 For a similar point regarding Audubon's time in New Orleans, Iannini, *Fatal Revolutions,* 253–279.

112 Like all of Audubon's studies, the work is actually mixed media, a highly inventive mixture of water-color, gouache, ink, and graphite is used to achieve a level of scientific detail not possible in traditional watercolor. See Marjorie Shelley, "Drawing Birds: Audubon's Artistic Practices," in *Audubon's Aviary: The Original Watercolors for* The Birds of America (New York: New-York Historical Society, 2012), 108–131.

113 The key study, devoted to Maria Martin but with evidence that relates to all members of the Bachman household, both enslaved and free, is Lindsay, *Maria Martin,* 9–10, 16, 46–112, 123. I also thank Dr. Lindsay for sharing with me archival documents that record Thomas's purchase by Bachman in May of 1831. South Carolina, Department of Archives & History, Series S213003, Vol. 5K, p. 430, May 20, 1831, "Marion, Jane to John Bachman, Bill of Sale for a Slave named Thomas."

114 He would also later write the published text for *Quadrupeds.* For correspondence that records Bachman sending Audubon specimens, Catherine L. Bachman, *John Bachman: The Pastor of St. John's Lutheran Church, Charleston* (Charleston: Walker, Evans & Cogswell Co., 1888), 98, 102.

115 Lindsay, *Maria Martin,* 47–48.

116 Lindsay, *Maria Martin,* 9, 197–198, 270 note 5.

117 Martin was also self-taught in basic botany and made ornithological observations in the field, see Lindsay, *Maria Martin,* 26, 46–112, 132, 134, 136, 138; Olson, "A Biographical Sketch," 26.

118 The letter is published in Bachman, *The Pastor*, 99–101, John Bachman to John James Audubon, Charleston, Dec 23, 1831, "Your visit to me gave me new life, induced me to go carefully over my favorite study, and made me and my family happy… Tell Henry Ward, that I will never make an attempt at painting, but that I am beginning to stuff birds, and *my man, Thomas, during my absence, has done the same.*" The same letter also records that Maria is beginning to paint birds, studying Audubon's compositions, "My *sister Maria, paints birds better every day; she fails only in setting them up.* Your book [i.e. Vol 1 of Birds] however, will soon be here, and *she will study the attitudes* of your birds. She is all enthusiasm, and I need not say to you that she is one of your warmest admirers… Am I not a bore to send you such a long letter to a tropical climate?" (Emphasis mine.) Cited in Lindsay, *Maria Martin*, 224 note 41.

119 This is indicated by the paint layer, which shows the completed stem of the plant underneath the semi-transparent body of the male bird. Oftentimes, however, Audubon would execute the bird first and the botanical or landscape would follow, sometimes with Audubon indicating its outlines in graphite. This suggests that Martin and Audubon were not both in Charleston at the time Martin executed the botanical watercolor. Shelley, "Drawing Birds," 111, compare the study for plate 321, 'Roseate Spoonbill'. The plant is identified on the original study in the New-York Historical Society in Maria Martin's hand as *Gordonia pubescens* rather than *Franklinia Alatamaha*. The former (*Gordonia pubescens*) follows the accepted name among British botanists, who disputed Bartram's identification of the plant as a distinct species. Audubon does not change the scientific name in the published *Ornithological Biographies*, John James Audubon, *Ornithological Biography*, vol. 2 (Edinburgh: Adam & Charles Black, 1834), 483–484. On the debate over the scientific name of Franklinia, Joel Fry, "Bartram's Tree: *Franklinia alatamaha*," in *The Attention of a Traveller: Essays on William Bartram's Travels and Legacy*, ed. Kathryn H. Braund (Tuscaloosa: University of Alabama Press, 2022), 73–114.

120 For other examples of Audubon's reliance on enslaved labor, see a letter relating to an expedition to Cole's Island, just south of Charleston, with Bachman and others. American Philosophical Society Collections, John James Audubon Papers, Mss.B.Au25, John James Audubon to Lucy B. Audubon, Nov. 7 1831, "Tomorrow….we go to Cole's Island…." This expedition is discussed by Audubon in published sources but without explicit mention of the enslaved laborers, John James Audubon, *Ornithological Biography*, vol. 3 (Edinburgh: Adam & Charles Black, 1835), 441–443.

121 Lindsay, *Maria Martin*, 71–73. It is interesting to note that the young Joseph Mason, who worked for Audubon between 1820–1822, severed ties with Audubon because he believed he had not been given sufficient credit, Lindsay, *Maria Martin*, 11.

122 It appears in the body of the textual description, see Audubon, *Ornithological*, 2: 483–484.

123 Blum, *Picturing Nature*, 112–113 notes that the narratives of the *Ornithological Biographies* are derived from the images, which take precedence, and she identifies his prioritization of images over text as a break from natural history tradition. The law required that Audubon deposit nine books free of charge in British public libraries, which would have cost him a fortune, Nobles, *John James Audubon*, 115–117.

124 Olson, "A Biographical Sketch," 26; Lindsay, *Maria Martin*, 84–86.

125 Lindsay, *Maria Martin*, 16; Bachman, *The Pastor*, 150.

126 Lindsay, *Maria Martin*, 9, 197–198, 270 note 5; Bachman, *The Pastor*, 356–357.

The study of the natural world, including animals,
plants, and minerals. Naturalists uncovered nature's order
by collecting and describing nature's diversity.

ACKNOWLEDGING INDIGENOUS HOMELANDS

The city of Philadelphia, including this museum, stands in the broader region of Lenapehoking,
the homeland of past, present, and future generations of the Lenape people. What is now Philadelphia
has long been a place of settlement and a site where people gathered for diplomacy and exchange.
Indigenous peoples from many nations continue these practices in Philadelphia today.

At the APS, the Center for Native American and Indigenous Research (CNAIR)
assists people in finding and utilizing our extensive collections by and about Indigenous peoples.
CNAIR promotes research that honors Indigenous knowledge, cultivates scholarship,
and strengthens languages and cultural traditions.

Illustrated Checklist

THROUGH WORDS AND IMAGES, William Bartram (1739–1823), Titian Ramsay Peale (1799–1885), and John James Audubon (1785–1851) captured nature's splendor. The careers of these natural historians spanned an exciting period in early American science.

Naturalists explored lands unfamiliar to Europeans. They engaged with new ideas and scientific approaches. They borrowed freely from art to express evocative ideas about nature.

At the same time, natural history was not without consequences. The work of these three naturalists claimed lands that belonged to Native nations. It promoted practices that ultimately led to environmental destruction. It endorsed racial hierarchies. Finally, they relied on the unacknowledged contributions of Native Americans and Africans.

Bartram, Peale, and Audubon's work is filled with injustice and innovation in equal measure. This exhibition invites you to engage with this complexity.

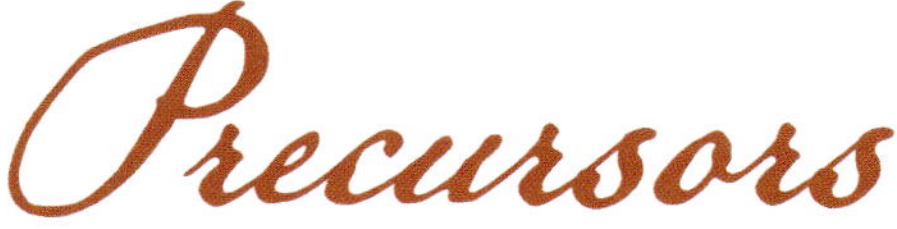

Precursors

EIGHTEENTH-CENTURY NATURAL HISTORY

Early American natural history was influenced by European science. Europeans offered diverse approaches to natural history. Some, like the Swedish botanist Carl Linnaeus (APS, 1769), focused on taxonomy, or identifying, naming, and cataloging new species. Linnaeus developed one of the most influential taxonomic systems, which classified plants according to their reproductive parts. French naturalist Georges-Louis Leclerc de Buffon (APS, 1768) thought Linnaeus' system was too simplistic. Buffon advocated detailed observation of living things in nature, including their behavior and relations to plants, animals, and habitats. He also hypothesized natural laws relating to ecology and evolution. Ideas about nature drawn from art also found their way into American natural history. Naturalists depicted the beauty of nature like painters and highlighted their emotional responses like poets.

FIG 12 PORTRAIT OF CARL LINNAEUS
J.M. Bernigeroth, 1749, Frontispiece for *Systema Naturae*, (6th ed.), Etching. Carolus Linnaeus. Etching by J. M. Bernigeroth, 1749. Wellcome Collection. Public Domain Mark. Source: Wellcome Collection.

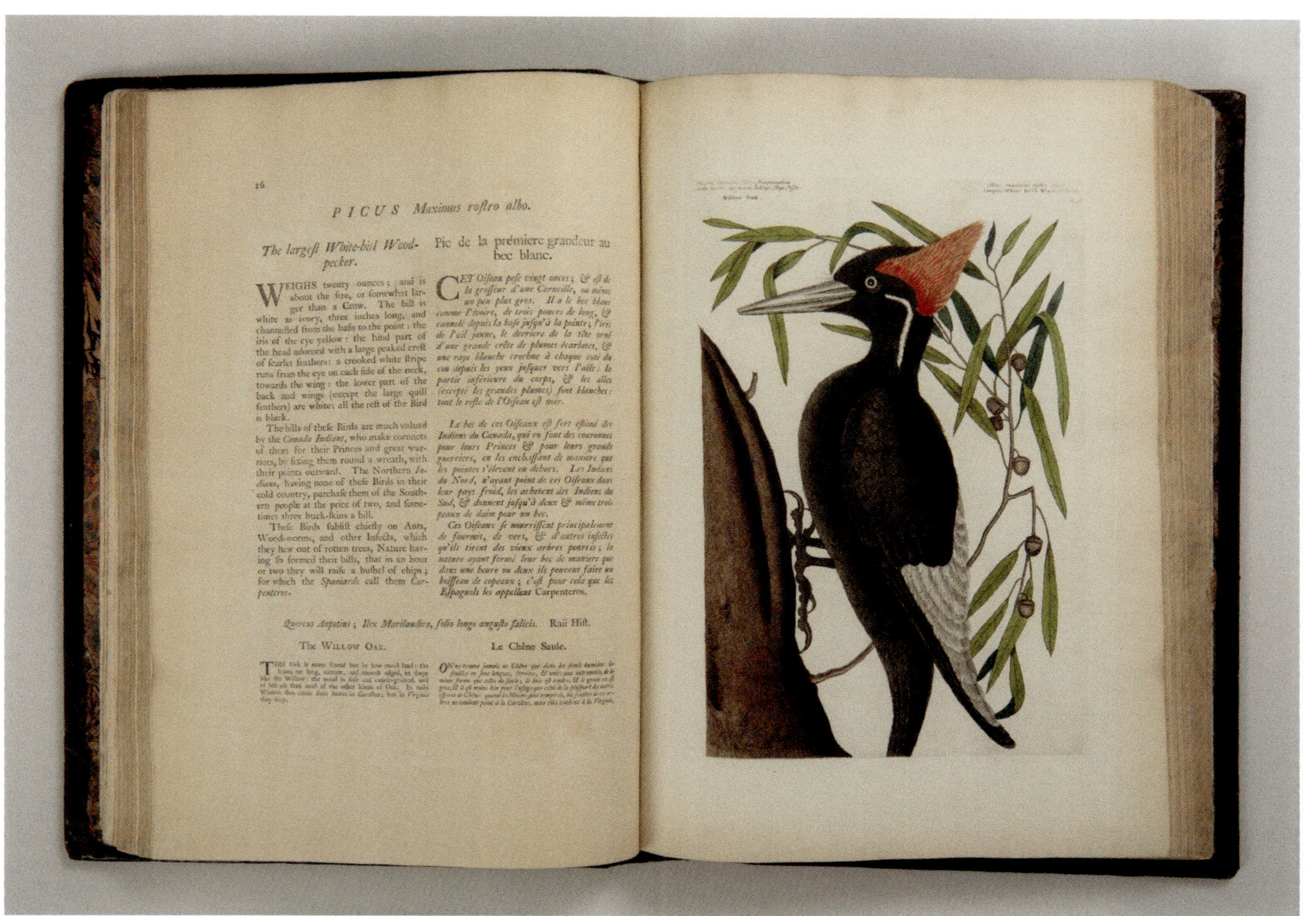

English naturalist Mark Catesby published the first significant illustrated natural history of the British colonies in the Americas, issued in parts between 1729 and 1747. Catesby was especially attentive to ecological relationships and relied on the knowledge of both Native Americans and enslaved Africans.

CAT 1
THE NATURAL HISTORY OF CAROLINA, FLORIDA, AND THE BAHAMA ISLANDS,
VOL. 1
Mark Catesby

London, 1771 (3rd edition)
Engraving with hand-coloring and letterpress in bound volume
APS.

TAXONOMY:

The scientific practice of naming and organizing natural things
into groups based on shared characteristics.

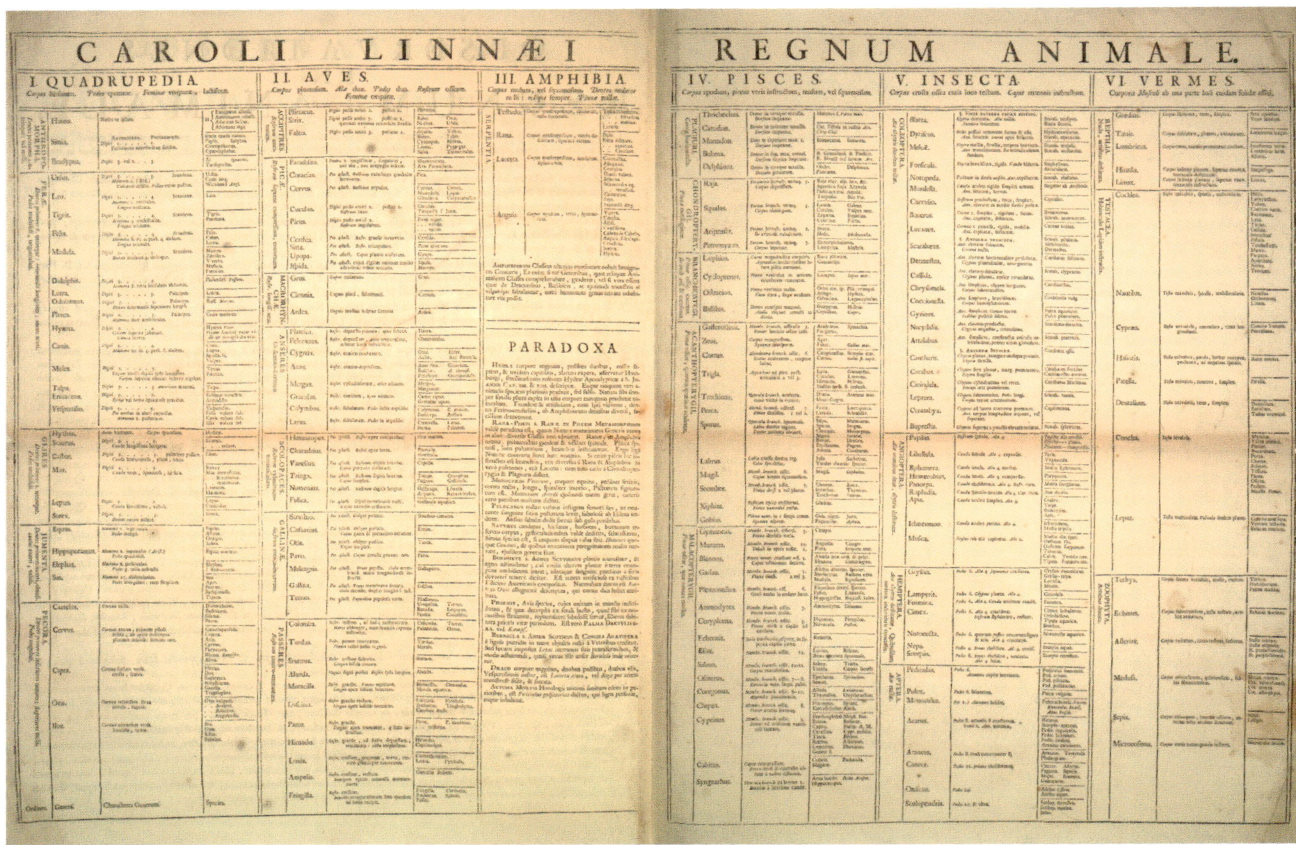

Linnaeus's publication laid out a five-tier hierarchical classification, or taxonomy. He divided the natural world into three kingdoms (animal, plant, and mineral), further subdivided into class, order, genus, and species. He also introduced his method for categorizing plants based on their reproductive parts. Linnaeus's system was the first comprehensive method for organizing all known natural forms.

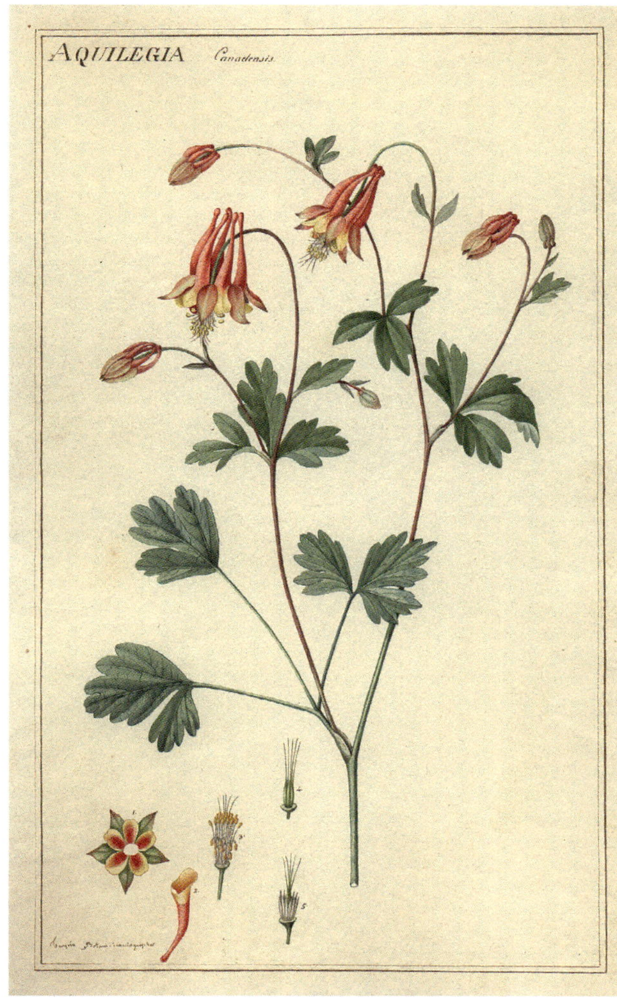

This is a typical example of a botanical specimen image, showing the columbine with no environmental context. It focuses on the physical qualities of the plant that are essential to taxonomy, showing the flower in various life stages and dissecting the flower, petals, and male and female reproductive parts at bottom left.

CAT 3
RED OR EASTERN COLUMBINE
Pierre Jean François Turpin
N.D.
Watercolor, graphite, and ink
APS.

CAT 4 (BELOW)
L'HISTOIRE NATURELLE DES OISEAUX, VOL. 1
Georges-Louis Leclerc de Buffon

Paris, 1771
Bound volume
APS. Presented by the author.

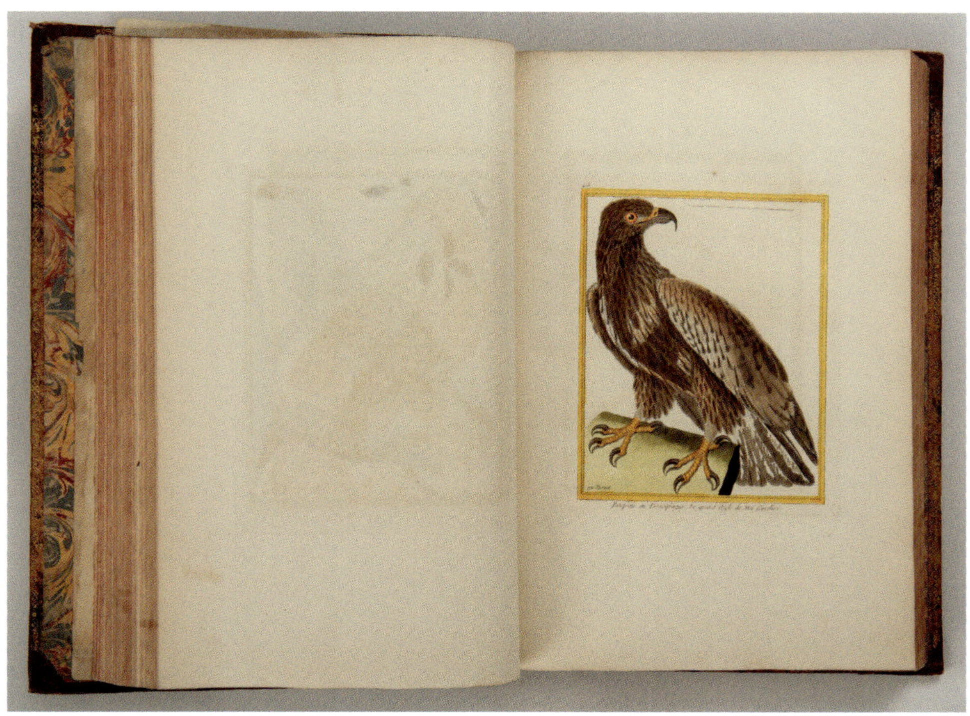

Buffon advocated firsthand observation of animals in their environments. He believed environments caused changes in humans, animals, and plants over time. Buffon argued that North America's inferior climate caused species to weaken or 'devolve.' Buffon used his false theories to claim that Europeans were superior to their colonial subjects.

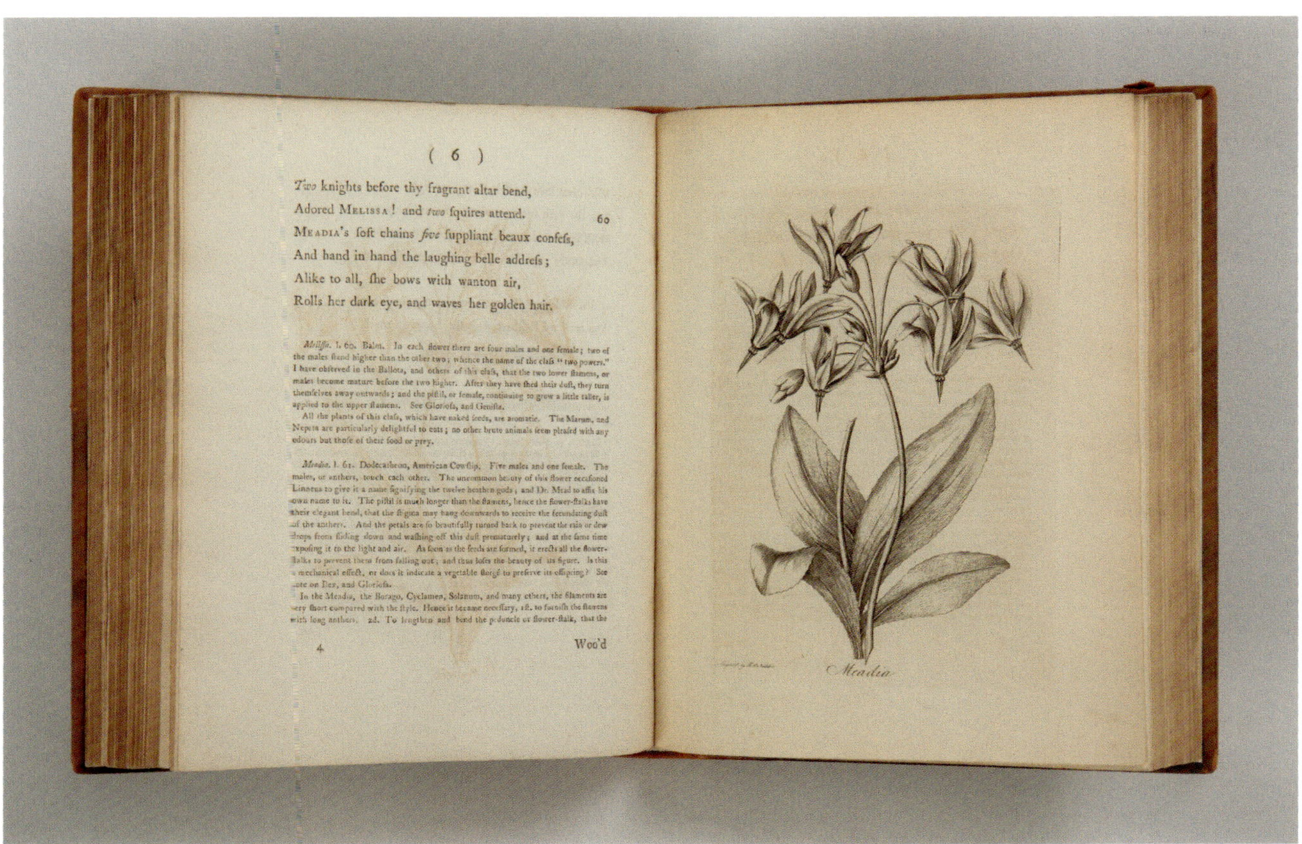

Erasmus Darwin, grandfather of Charles, communicated scientific ideas through poetry that described imaginative botanical 'romances.' Erasmus sought to delight as well as to inform, demonstrating the frequent fusion of art and science in the 18th century.

CAT 5
THE BOTANIC GARDEN . . .
PART II, THE LOVES OF
THE PLANTS
Erasmus Darwin

London, 1791
Bound volume
APS. Presented by the Michaux Fund.

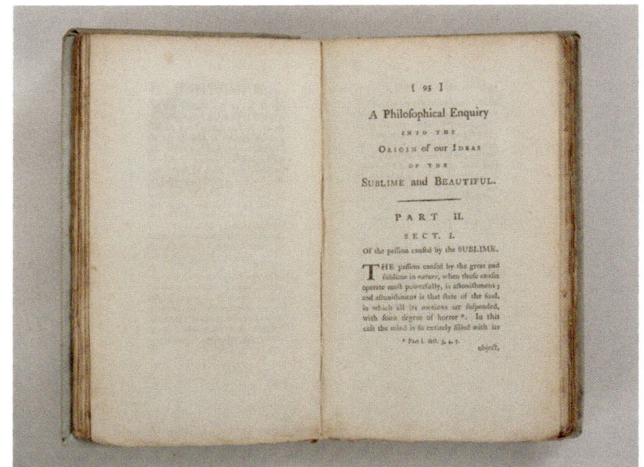

CAT 6
A PHILOSOPHICAL
ENQUIRY INTO THE
ORIGINS OF IDEAS OF THE
SUBLIME AND BEAUTIFUL
Edmund Burke

London, 1787
Bound volume
APS. Presented by the descendants of Benjamin Vaughan through Mrs. George Gibson in 1991.

Edmund Burke used the term 'sublime' to describe aspects of the natural world that seemed vast, infinite, and mysterious. Burke argued these natural phenomena were so awe-inspiring that they overwhelmed the rational mind. Science alone could not explain these phenomena, requiring more artistic modes of expression.

NATIVE AMERICAN EXPERTS

Native American experts had, and continue to have, deep knowledge of North America's biodiversity and ecology. Euro-American scientists in the 18th and 19th centuries used this knowledge when developing their ideas. Although they relied heavily on Native expertise in their work, they rarely acknowledged the contributions of specific Native individuals. Moreover they made racist claims that Native Americans were socially and intellectually inferior to people of European descent. Appropriating Native expertise was another way that Euro-Americans could assert both dominion over the American landscape and their own intellectual authority.

Lenape names for natural landmarks in Pennsylvania are descriptive and reveal detailed ecological knowledge. For example, the Lenape place name 'Nolamattink'—an area near Nazareth, Pennsylvania—means, "the place where the silkworms spring up," or mount their cocoons. The silkworms attached their cocoons to local mulberry trees.

CAT 7
NAMES WHICH THE LENNI LENAPE . . . HAD GIVEN TO RIVERS, STREAMS, PLACES, ETC.
John Heckewelder

1822
Manuscript
APS. Presented by John Heckewelder.

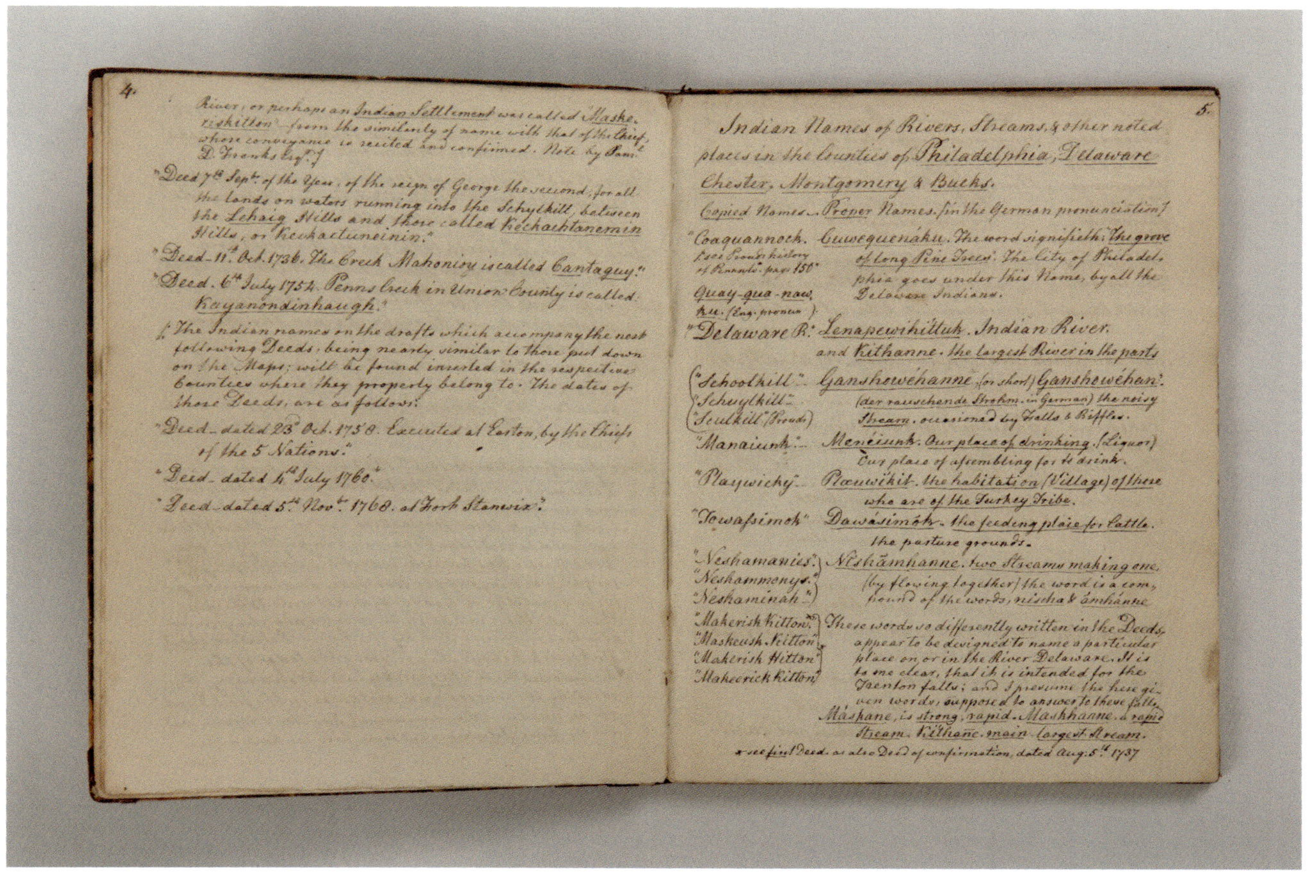

Benjamin Smith Barton, a student of William Bartram, described Native Americans' knowledge of botany, zoology, and geology as highly valuable and reliable. However, he refused to give Native American experts the same status as Euro-Americans.

CAT 8
A DISCOURSE ON SOME OF THE PRINCIPAL DESIDERATA IN NATURAL HISTORY
Benjamin Smith Barton

Philadelphia, 1807
Ink on paper bound volume
APS. Presented by the author, 21 Aug 1807.

A

DISCOURSE

ON

SOME OF THE PRINCIPAL DESIDERATA

IN

NATURAL HISTORY,

AND ON

THE BEST MEANS OF PROMOTING THE STUDY

OF THIS

SCIENCE,

IN THE UNITED-STATES.

READ BEFORE THE PHILADELPHIA LINNEAN SOCIETY,
ON THE TENTH OF JUNE, 1807.

BY BENJAMIN SMITH BARTON, M. D.,
PRESIDENT OF THE SOCIETY; ONE OF THE VICE-PRESIDENTS OF THE
AMERICAN PHILOSOPHICAL SOCIETY; AND PROFESSOR OF
MATERIA MEDICA, NATURAL HISTORY AND BOTANY,
IN THE UNIVERSITY OF PENNSYLVANIA.

PHILADELPHIA:
PRINTED BY DENHAM & TOWN,
No. 278, South Second-street.
1807.

William Bartram

THE BORDERS OF A 'NEW' WORLD

William Bartram (APS, 1768) witnessed the emergence of both the American nation and its scientific community. His father, John Bartram, ran a plant nursery outside Philadelphia and helped found the American Philosophical Society in 1743. William's major published work, the *Travels* (c. 1791), was one of the new American republic's first natural history books.

William Bartram's work shows us that contemporary views of the American landscape were complex. On the one hand, he describes nature as a source of creative, intellectual, and spiritual inspiration. However, William also presents nature as a source of valuable resources, ripe for development.

William also struggled to articulate a clear position on the competing claims to this landscape. He advocated the equal rights of Native Americans, but ultimately accepted the republic's occupation of their homelands.

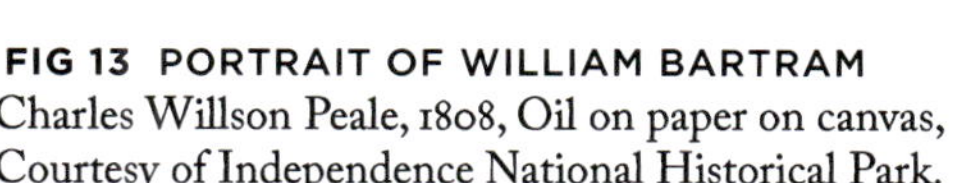

FIG 13 PORTRAIT OF WILLIAM BARTRAM
Charles Willson Peale, 1808, Oil on paper on canvas,
Courtesy of Independence National Historical Park.

BARTRAM'S *TRAVELS*

William Bartram's *Travels* records his 1773–1777 journey through the American Southeast. Bartram's text combined multiple ways of understanding nature from science and literature. Using the tools of the taxonomist, he cataloged the American plants and animals that he encountered. These taxonomic lists advanced science but also highlighted potentially profitable natural resources. In addition to cataloging individual plants and animals, William also described the relations between them and their environments— what we would call ecology. Finally, he included poetic passages that described nature as something beyond human comprehension–mysterious, vast, and divine. The *Travels* shows us Bartram's many visions of nature.

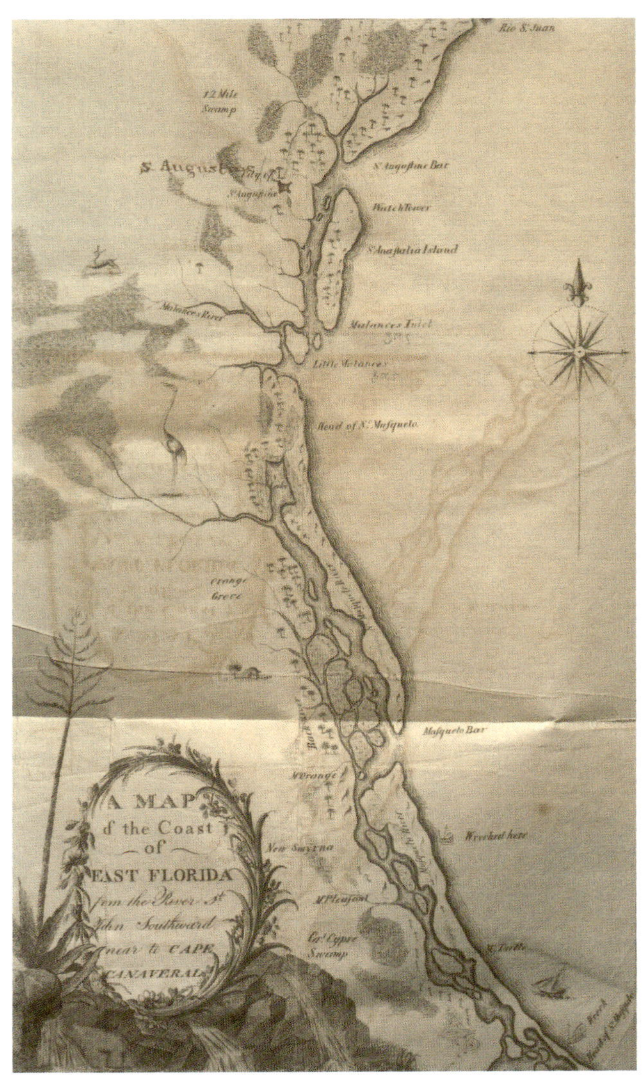

FIG 14 A MAP OF THE COAST OF EAST FLORIDA
William Bartram, 1791, In *Travels through North and South Carolina, Georgia, East and West Florida*, Engraving and letterpress in bound volume, APS.

William Bartram produced this map for the *Travels*, showing part of the region he explored in Florida.

CAT 9
A DESCRIPTION OF EAST-FLORIDA: WITH A JOURNAL . . . BY JOHN BARTRAM OF PHILADELPHIA . . .
William Stork

London, 1769
Bound volume
APS. Presented by Michaux Fund.

In 1765, before his *Travels* journey, William Bartram traveled to Georgia and Florida with his father John Bartram, who had been appointed King's Botanist. John identified useful plants and territory for his royal patron, and he paid attention to plant ecology. William added new ways of understanding nature in his *Travels*.

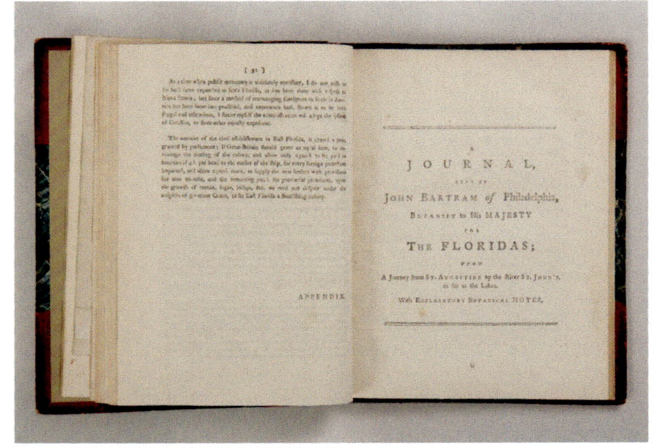

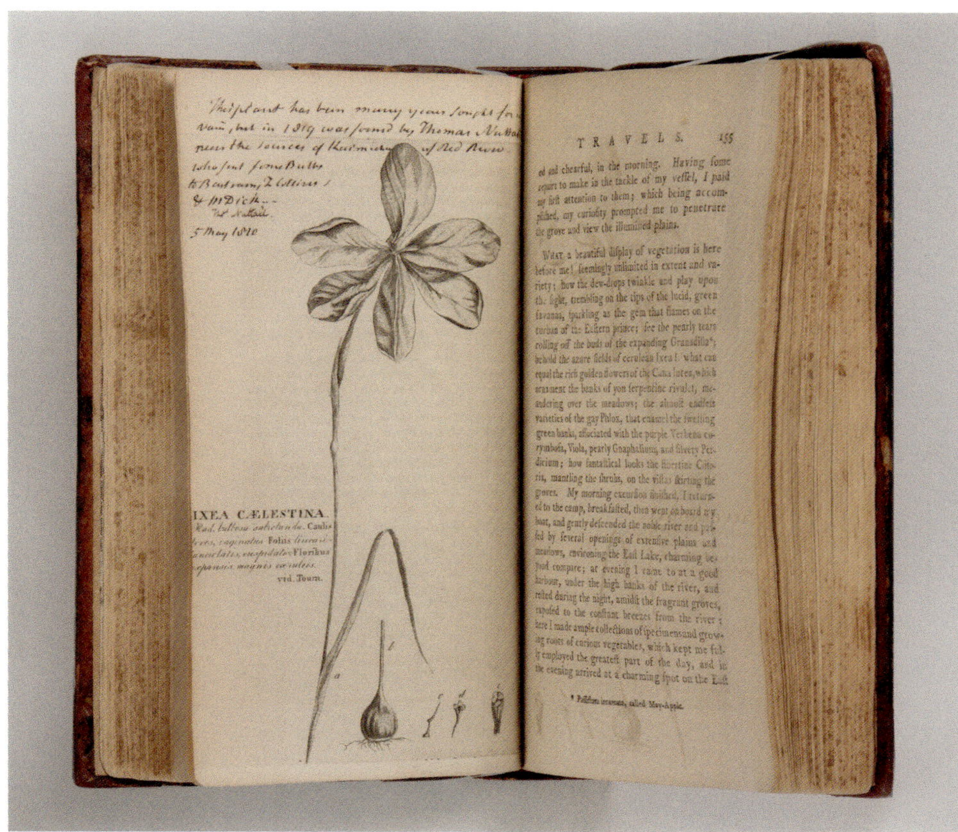

Bartram's *Travels* included standard descriptions and uses of American flora and fauna. These were important to American scientists who wanted to strengthen their claims to the national landscape. However, the *Travels* distinguished itself from most natural history texts by including reflective language and underlining nature's splendor.

CAT 10 (ABOVE)
TRAVELS THROUGH . . . EAST AND WEST FLORIDA . . .
William Bartram

Philadelphia, 1791
Handwritten ink and printed ink in bound volume
APS. Gift of William Bartram, 1792.

CAT 11 (RIGHT)
DRAFT MANUSCRIPT OF THE TRAVELS
William Bartram

Philadelphia, c. 1777-1786
Manuscript
The Collection of the Historical Society of Pennsylvania.

Bartram's initial draft of the *Travels* focused on the unspoiled beauty of what he encountered and was skeptical of American expansion. This page, for example, which is marked by a box in brown ink indicating that it was deleted from the final version, focused on ecological connections as part of a divine plan.

The study of living things, their relationships to each other,
and their environments.

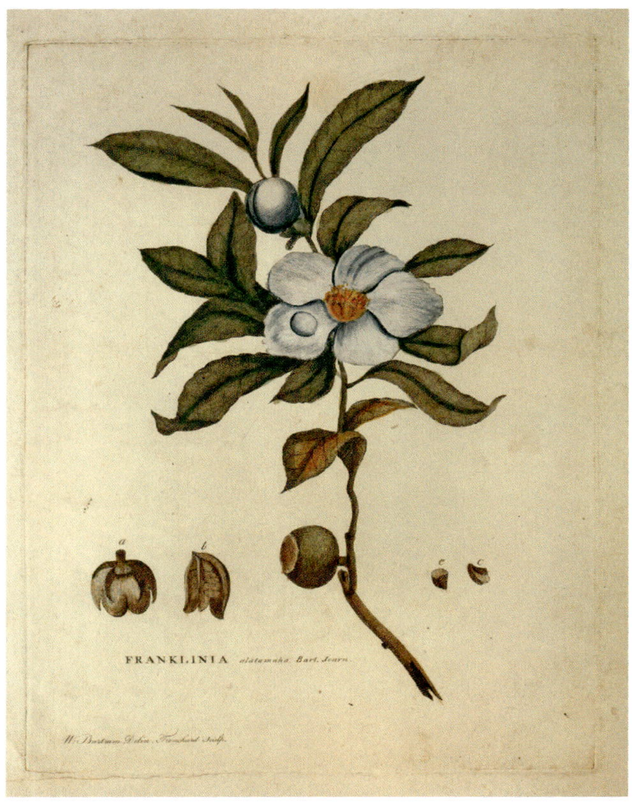

Bartram's botany was not just about science; he also identified unusual plants that might turn a profit. He and his father spotted a Franklin tree on the Altamaha River in 1765. William later brought back seeds of this rare and beautiful plant to cultivate in the Bartram family nursery, naming it after Benjamin Franklin.

CAT 12
FRANKLIN TREE
William Bartram

Philadelphia, N.D.
Engraving with printed ink
and hand-coloring
APS.

Bartram's journeys supplied plants to sell in the family nursery. This list records a shipment of plants for an unknown customer. Number four is the Franklin tree with instructions to plant the tree in loose soil. It is now extinct in the wild and all living trees descend from the plants Bartram cultivated.

CAT 13
PLANT LISTS
William Bartram

Philadelphia, 1780s
Manuscript
APS.

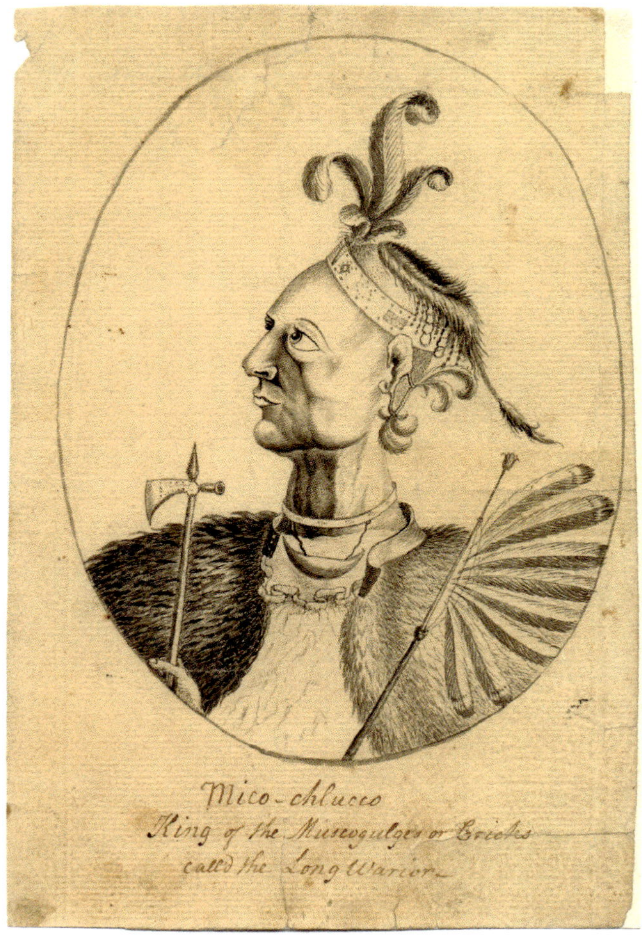

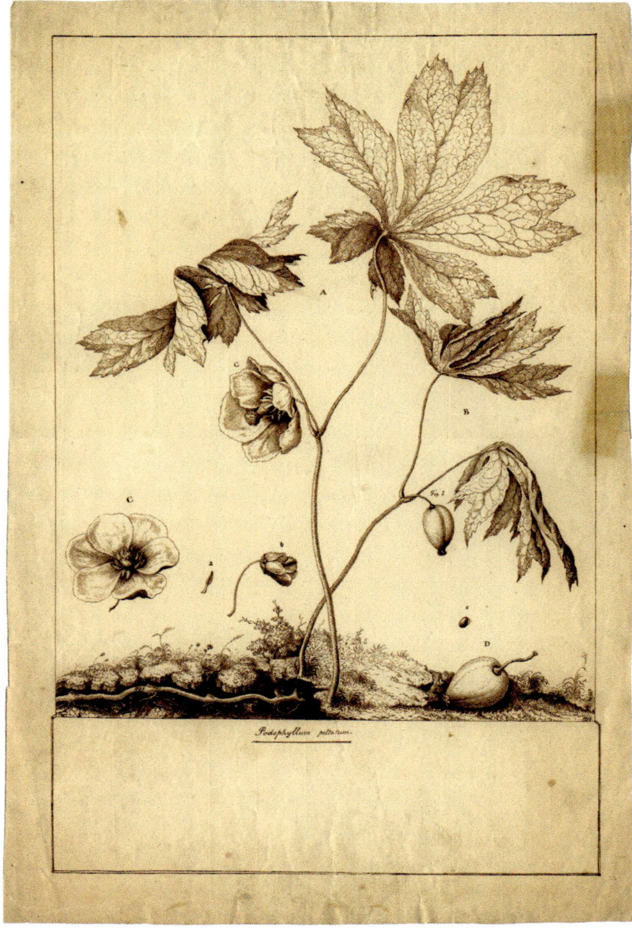

Mico-Chlucco, a major Seminole ruler, is depicted in his regalia, the feathers in his headband indicating his high status. The Seminole are a Muscogee group in Florida that gradually gained a separate identity during the late eighteenth century. Mico-Chlucco's portrait served as the frontispiece for the *Travels*. Bartram emphasized the equal rights of Native Americans in the *Travels*, but later embraced a government plan for forced assimilation that sought to erase their cultures.

Bartram learned about medical botany from the Cherokee, Muscogee, and Seminole. His drawing of a mayapple pays special attention to its roots, which southeastern Native Americans used for various medicinal treatments. Bartram personally experienced mayapple's curative effects. A compound derived from mayapple is used today in certain chemotherapies.

CAT 14
MICO-CHLUCCO, KING OF THE MUSCOGULGES
William Bartram

Philadelphia, N.D.
Watercolor, graphite, and ink on paper
APS.

CAT 15
MAYAPPLE
William Bartram

Philadelphia, 1793
Ink on paper
APS.

Kingsess Decemb.r 29th 1792.

My ingenious worthy Friend

I recieved thy Letter this week dated 1.t Decem.br together with the Book, which are exceedingly pleasing to me. I shall peruse this fine Volume, and be carefull, untill I return it to Thee.

I am greatly pleased to see a Figure and description of the Asiatic Plotus particularly, because it induces me to believe that the Snake Bird of Carolina and Florida, is a Species of that Genus, and not widely different. The may see a short account of the Snake Bird in my Journal of Travels

I wish it were in my Power to answer, to thy satisfaction, thy usefull Queries concering the Materia Medica of the Creeks and othe Nations of Red men; a perfect knowledge of which would be very interesting to Human Society.

I think I have observed that they have few Complaints, or bodily ailments in comparison to what Afflict the inhabitants of the Old Continent, or the White people of this.

Yet sometimes they have Mortal diseases, very calamitous even to the depopulation of Towns, & almost extermination of whole Tribes, but these Awfull visitations happen but seldom, but when they do come it baffles the skill of their Physicians Then they resign themselves to their fate. These scourges of Mortality seem to fall greivously on their Children, in Hooping Coughs, & Sore Throat which they have no remidy for, as I was informed. A greivous distemper calld Pleurisy in the Head which is contagious and sweeps away the Adult particularly Hearty Men The Fall fever or inflamatory Fever sometimes destroy many of their Adults. & Small Pox destroys all in its way, few escaping with their Life, & those miserably disfigurd. I bilive the pestilential, Or putred contagious fevers, by which I mean the Yellow, & Spotted Fevers have not yet reach't them

The Ague is very common

I do not know what particular remedy they use against each disorder, but I believe that most if not all their remedies are Vegetables applyed in various ways, particularly, Emetics, Cathartics, Sudorificks, & Diureticks. The infusion or decoction of the leaves & tender young shoots of Ilex Capine is perhaps the most powerful & effeacious vegetable Diuretick yet known, for its effects are almost instantanious after the Draught, which I have experienced often. This famous decoction (calld Black Drink) is consider'd rather as a preventative, than a Medicine, & I do not recollect that it is properly amongst their Physick Plants, yet held in divine estimation, suppos'd to be ordaind for the preservation of their health The Indian Physicians, injoin the patient strictly to regimen, during their attendance on them And the Stove or Sweat houses are in constant practice.
They use Phlebotomy, by Scratching & I bilive Cuping, or at least sucking the blood out of the scarifieations.
The Lower Cricks, Siminoles in particular, live perhaps in the lowest, & most inundated country on the Globe, yet appear to be healthyer, & have fewer disorder than the Upper Creeks, whither they derive these advantages from the natural food & Air of the Country, I know not

AN ECOLOGICAL VISION

Throughout his life, William Bartram used many different approaches to representing nature. Sometimes he showed plants and animals isolated from context. In other images, he placed living things in wider environments, or depicted entire ecosystems. Bartram's drawings also introduced elements that emphasized nature's awesome vastness or its infinite variety, qualities that he saw as evidence of nature's divine maker. Through his images, Bartram invited his viewers to approach the natural world in many ways, whether through objective science, appreciation of its beauty, or contemplation.

William Bartram

N.D.
Ink on paper
APS.

"… how is the mind agitated and bewildered, at being thus, as it were, placed on the borders of a new world!"

(William Bartram, *Travels*, 1791)

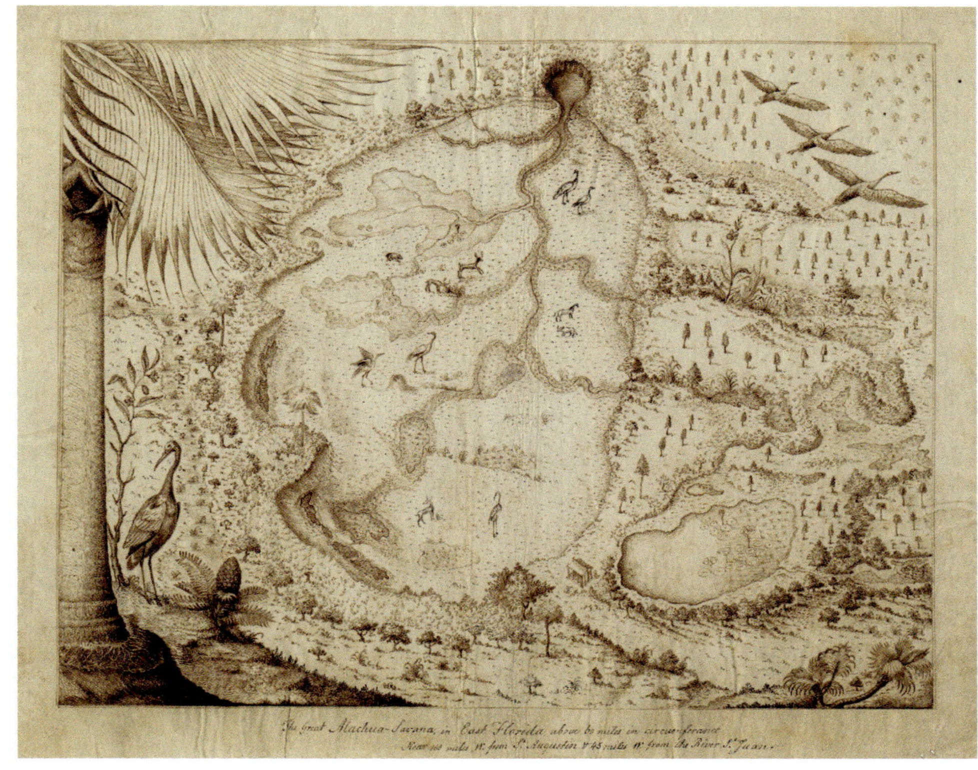

William Bartram's 1774 visit to the Alachua Savanna in Florida showed him a new aspect of nature. He was awestruck by both the variety of flora and fauna and the interrelationships between them. Bartram sought to capture all this beauty and complexity in his image of the savanna's entire ecosystem.

Despite his awe, Bartram nonetheless described the savanna's potential for settlement only two months later. He might even have known about the designs of a local planter to take the area from the Seminole of Cuscowilla, the village represented in the bottom corner of his image.

While Bartram emphasized the limits of taxonomy, his patrons did not all share this skepticism. In this watercolor commissioned by his student and collaborator Benjamin Smith Barton, Bartram adopts the standard presentation of Linnaean-style specimen images, highlighting the flower's visible form and the arrangement of its reproductive parts.

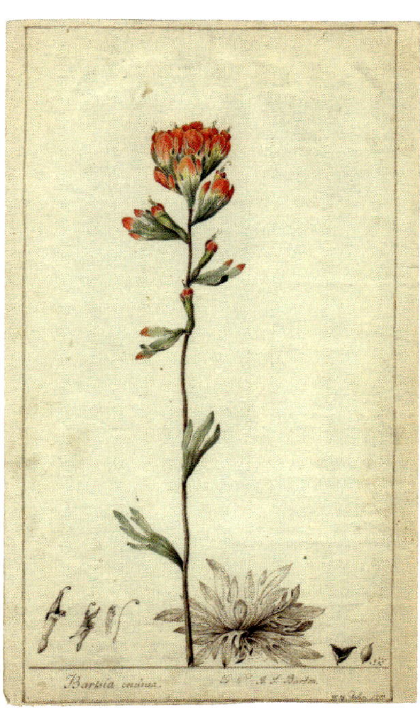

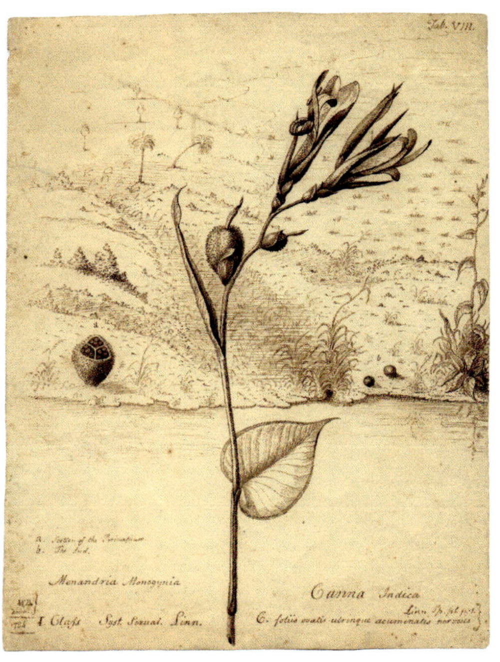

CAT 18 (LEFT)
RED BARTSIA
William Bartram

Philadelphia, 1801
Watercolor and ink
on paper
APS.

CAT 19 (RIGHT)
**RED CANNA OR
INDIAN SHOT**
William Bartram

Philadelphia, 1784
Ink on paper
APS.

Bartram depicted the Red canna flower twice. The large-scale image in the foreground is a standard specimen image, focusing on the plant itself. But in the second image, to the right, the plant is situated in a vast landscape, suggesting the complexity of its ecosystem and nature itself.

CAT 20
WOOD TURTLE
William Bartram

Philadelphia, N.D.
Ink on paper
APS.

The back of this drawing includes a lengthy description by Bartram comparing the turtle's shell to a beautifully carved sculpture. The fragmentary presentation of the turtle—which smiles endearingly despite having lost its head—suggests that human knowledge of living animals will always be incomplete.

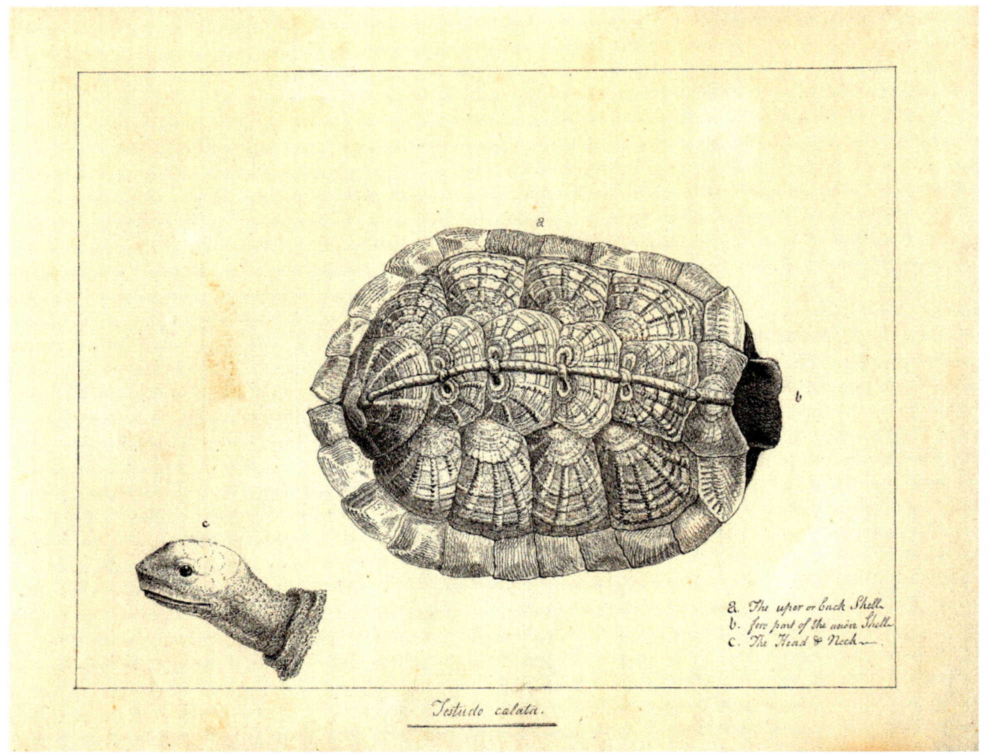

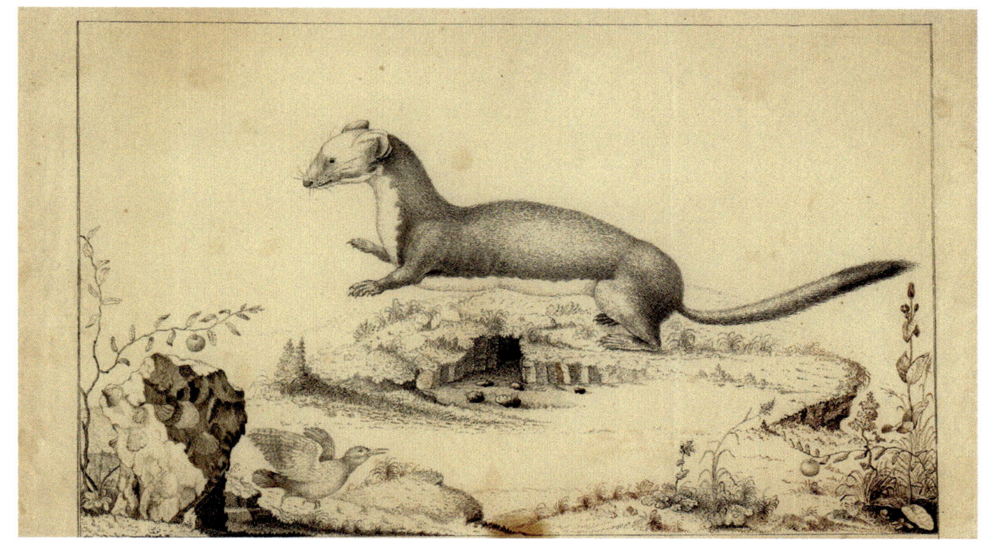

CAT 21
ONE OF OUR WEASEL
William Bartram
Philadelphia, N.D.
Graphite and ink on paper
APS.

The weasel is shown in pursuit of a bird. In the left foreground is a rock containing fossilized shells, which references the earth's ancient history. This image proposes that animals can only be fully understood within a complex web of relations and over time.

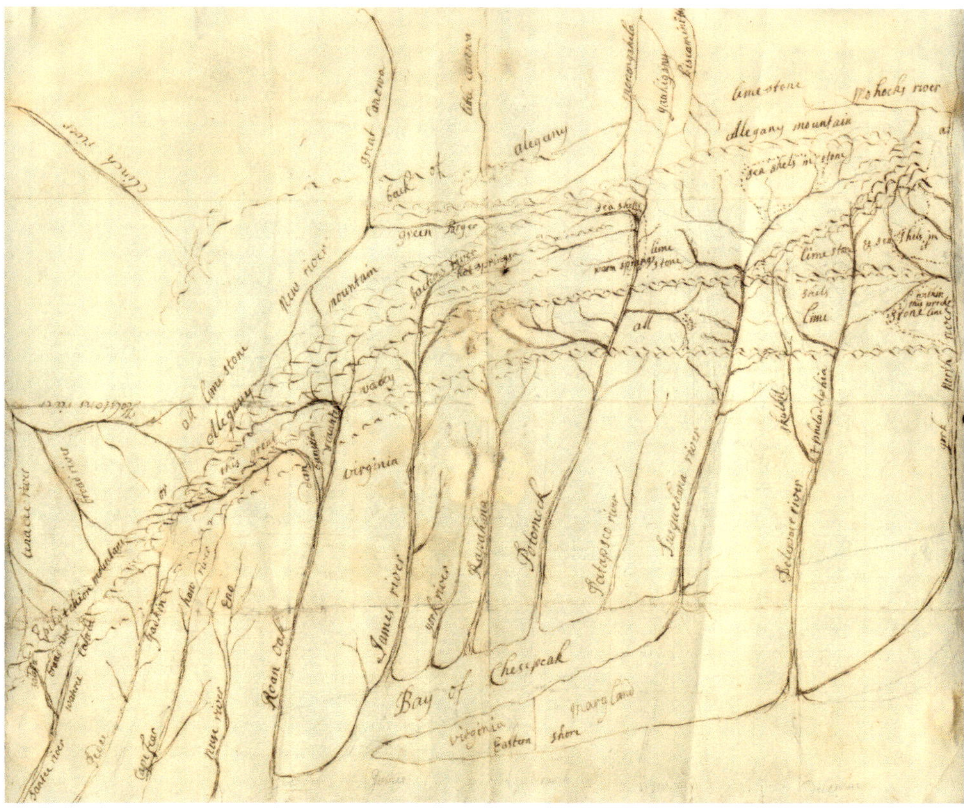

CAT 22
MIDDLE ATLANTIC STATES, SHOWING RIVERS AND MOUNTAINS AND LOCATION OF SEA SHELLS ON THE TOPS OF THE MOUNTAINS
John Bartram

1750s
Graphite and ink on paper
APS. Gift of Benjamin Franklin.

John Bartram identified the locations of fossilized seashells on mountaintops in this map of the Allegheny Mountains region. The shells suggested that the mountains had once been underwater, and that the earth's surface had changed over time. William Bartram references similar ideas in his *Weasel* drawing.

THE HIERARCHY OF NATURE

William Bartram believed nature offered a model for a more equal American society. European taxonomists described the natural world as rigidly hierarchical, with humans at the top, separated from animals and plants. In contrast, Bartram resisted the idea of 'natural' hierarchy by blurring these distinctions. He emphasized the human-like reason of animals, and the ability of certain plants to move like animals. Moreover, he saw the peaceful interactions between living things as evidence that harmony and equality were universal ideals. Bartram did not apply these ideals evenly. He enslaved people in the 1760s and 1770s, before adopting an abolitionist position later in life as public sentiment changed.

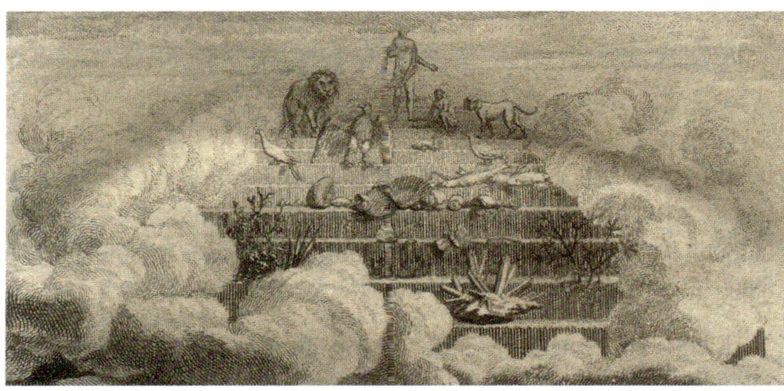

FIG 15 THE GREAT CHAIN OF BEING
Charles Bonnet, c. 1779–1783, *Oeuvres D'Histoire Naturelle et de Philosophie*, vol 4, Engraving, APS.

Most 18th-century naturalists imagined the natural world as hierarchical, with human beings placed at the top.

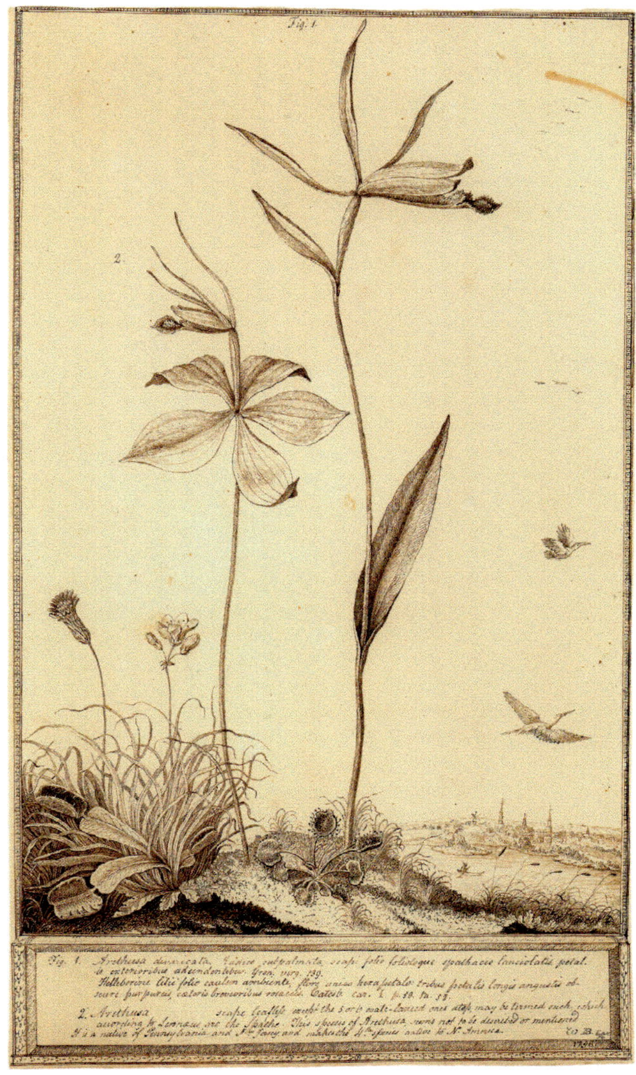

In this view from Bartram's Garden, the coexistence of two types of orchid with the carnivorous Venus flytrap in the foreground attests to the harmony across diverse species in nature. The garden is compared to the similarly diverse city of Philadelphia in the backdrop, suggesting nature offers a model for human society.

CAT 23
ROSEBUD ORCHID, WHORLED POGONIA OR PURPLE FIVE-LEAF ORCHID, VENUS FLYTRAP, AND ROUND-LEAF SUNDEW, WITH PHILADELPHIA (?) IN BACKGROUND . . .
William Bartram
Philadelphia, 1796
Ink on paper
APS.

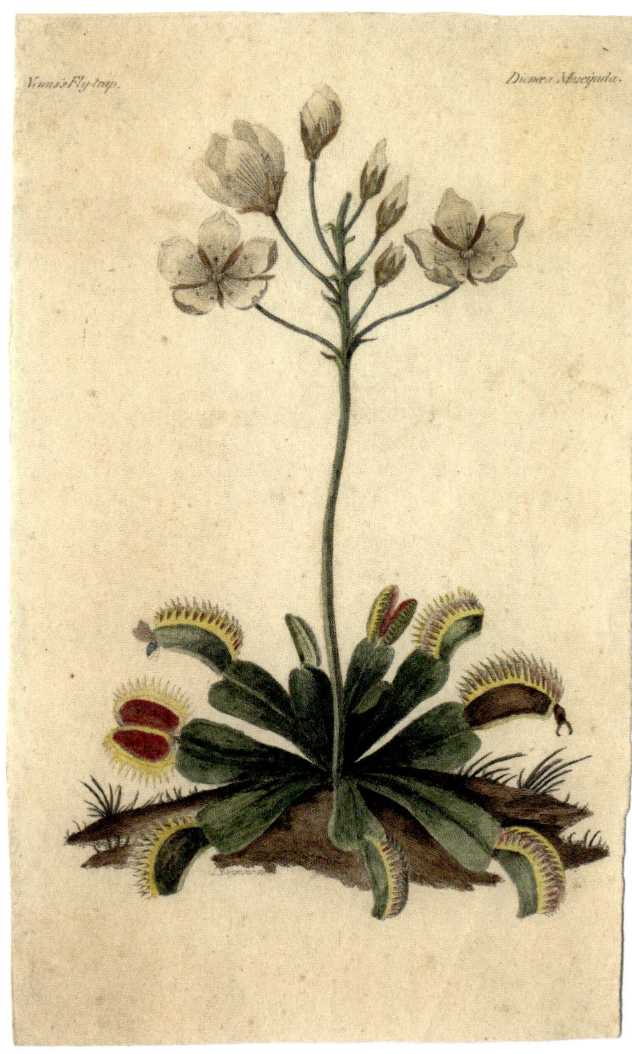

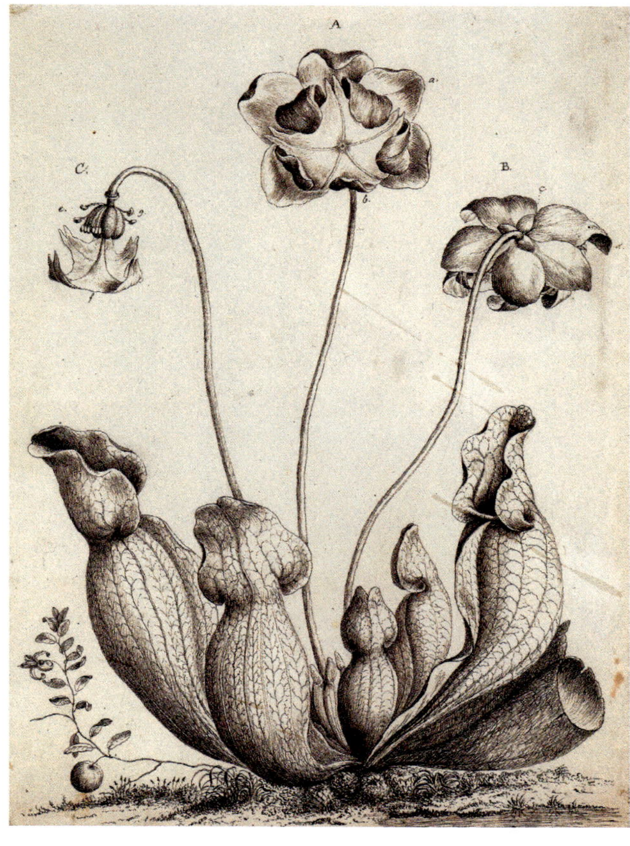

Bartram delighted in the Venus flytrap, whose leaves trap unsuspecting insects to become food for the plant. The flytrap supported Bartram's belief that scientists were incorrect when they claimed plants were less complex than animals. Both had the ability to move and catch prey.

Pitcher plants lure insects into cavities formed by leaves. Bartram was not certain whether the insects were food for the plants. The liveliness of his drawing invests this carnivorous plant with a sense of animal-like animation nonetheless.

CAT 24
VENUS FLYTRAP
I.H. Seymour, after
William Bartram

N.D.
Hand-colored engraving
APS.

CAT 25
PURPLE PITCHER PLANT
William Bartram

Philadelphia, N.D.
Ink on paper
APS.

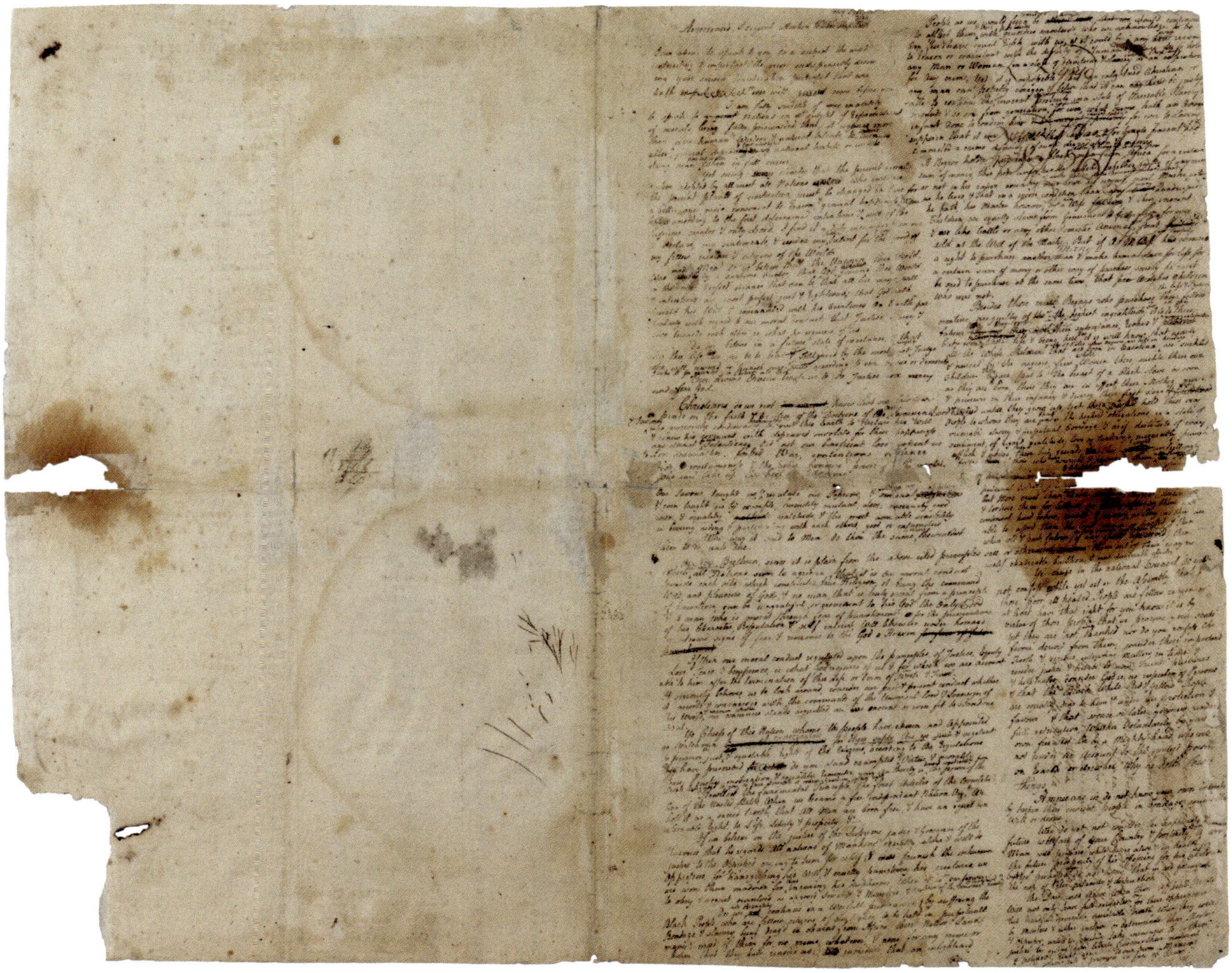

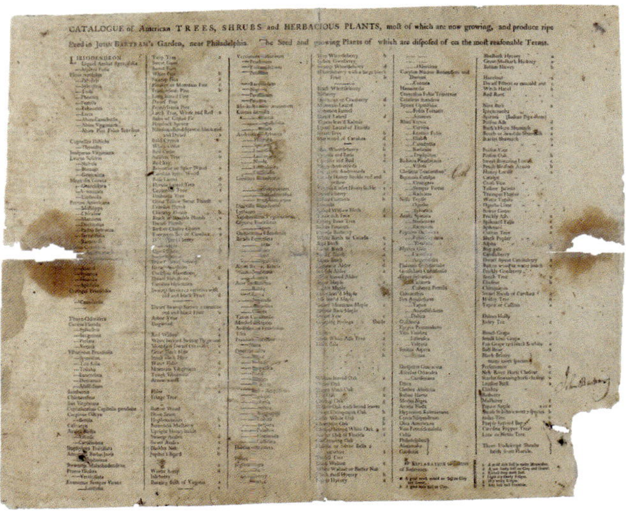

CAT 26
ABOVE: **ANTISLAVERY TREATISE**
William Bartram

c. 1790

LEFT: Written on John Bartram Jr. and William Bartram, *Catalogue of . . . plants . . . in John Bartram's Garden . . .* (c. 1783)
Broadsheet with manuscript notes
The Collection of the Historical Society of Pennsylvania.

Bartram wrote an antislavery treatise on the back of a catalog for his family's nursery around 1790. He adopted an abolitionist stance late in life and in response to Philadelphia Quakers' increasing support for the movement.

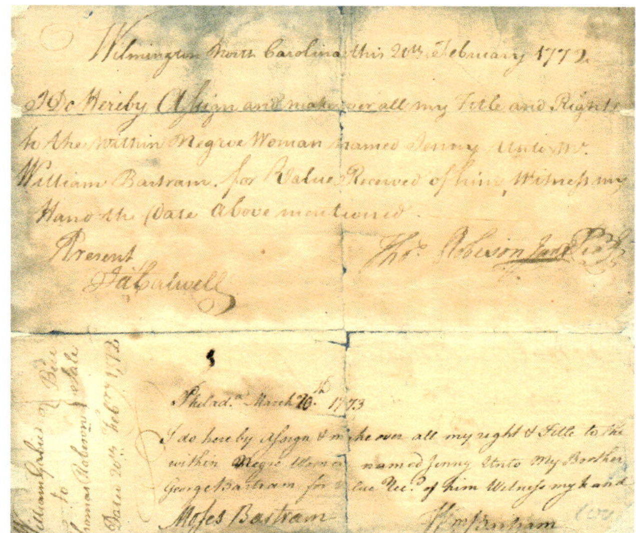

CAT 27
BILL OF SALE

1772–1773
Manuscript
APS.

This bill of sale records Bartram's enslavement of a woman called Jenny between 1772 and 1773. His brother Moses and brother-in-law George sold her on Bartram's behalf at the start of his Southeastern travels, and he may have used the money to partially fund his journey.

CAT 28
JOHN ELLIS LETTERS, 1753–1771

Meriel Nevill Watt (attr.), after John Ellis

N.D.
Manuscript
APS.

An unnamed enslaved man was credited with discovering a new species of star anise that John Bartram spotted a few months later. People of African descent often contributed unacknowledged, skilled work. William also learned about botany from free African Creoles during his time in Louisiana.

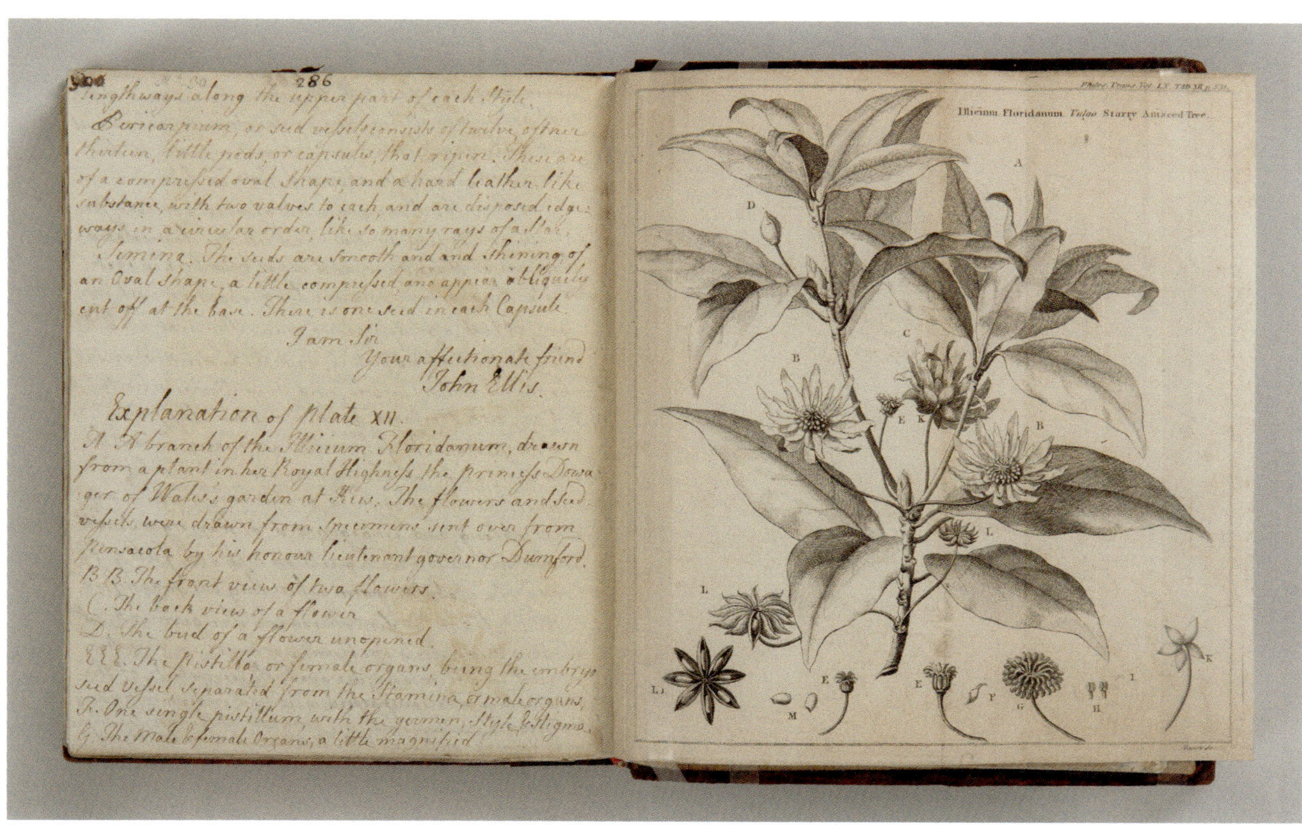

William Bartram instructed many significant early American naturalists, including Benjamin Smith Barton (APS, 1789), Alexander Wilson (APS, 1813), and Thomas Say (APS, 1817). Bartram's work influenced this generation of naturalists by fostering an emphasis on nature as an element of American identity. He also set an example for the importance of field study and the accurate observation of living things in their natural environments. The literary aspects of his work were not universally accepted, however, as scientists like Say increasingly emphasized professional scientific description and taxonomy.

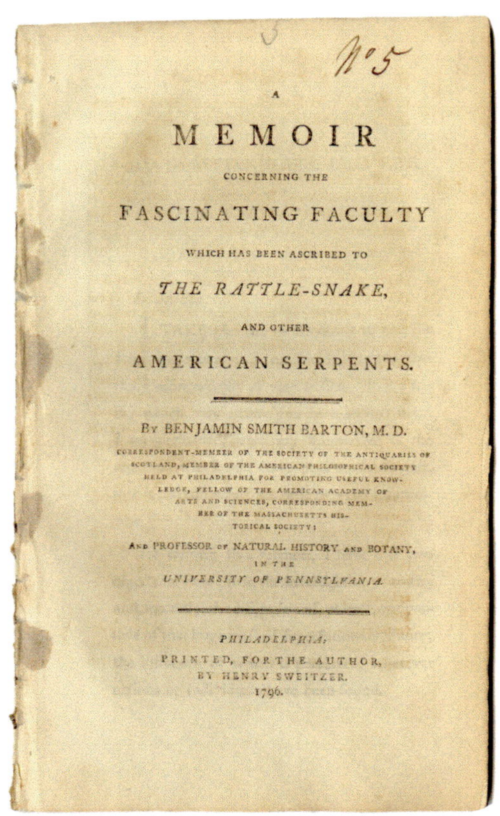

Benjamin Smith Barton was William Bartram's protégé. In this essay Barton relies heavily on Indigenous expertise to disprove the belief that rattlesnakes had the power to enchant, or 'fascinate,' their prey. This belief was common among European scientists including Linnaeus.

CAT 29
A MEMOIR CONCERNING THE FASCINATING FACULTY . . . ASCRIBED TO THE RATTLE-SNAKE . . .
Benjamin Smith Barton

Philadelphia, 1796
Bound volume
APS.

CAT 30 (BELOW)
RATTLESNAKE SKELETON
Attributed to Benjamin Henry Latrobe

N.D.
Watercolor, graphite, and glaze on paper
APS.

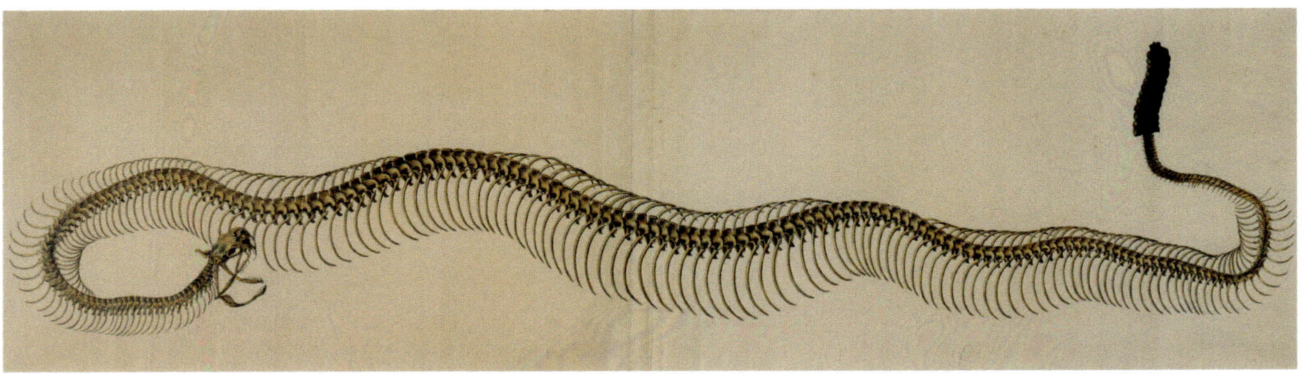

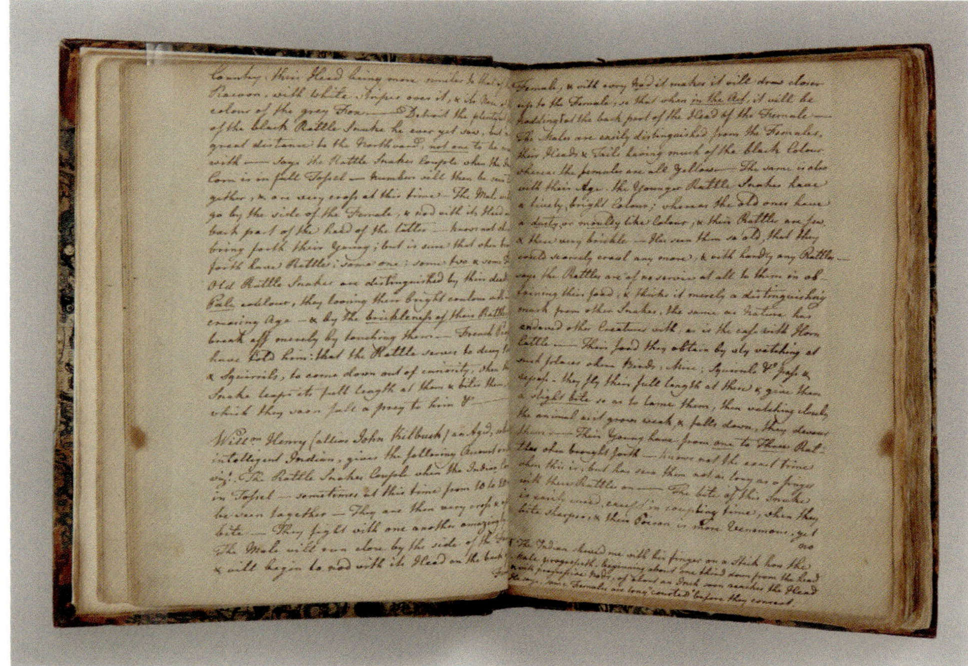

Barton relied heavily on Native American expertise in his rattlesnake treatise, some of it acquired secondhand via Moravian missionary John Heckewelder. Neither Barton nor Heckewelder named their Indigenous sources. However, in a later letter, Heckewelder detailed expert observations on rattlesnake hunting made by a Lenape man called John Killbuck Jr.

Benjamin Smith Barton collected natural history images as part of his scientific studies, including these two striking rattlesnake studies by the architect and engineer Benjamin Henry Latrobe. The clarity and precision of Latrobe's studies demonstrates his commitment to accuracy and close observation, hallmarks of American natural history.

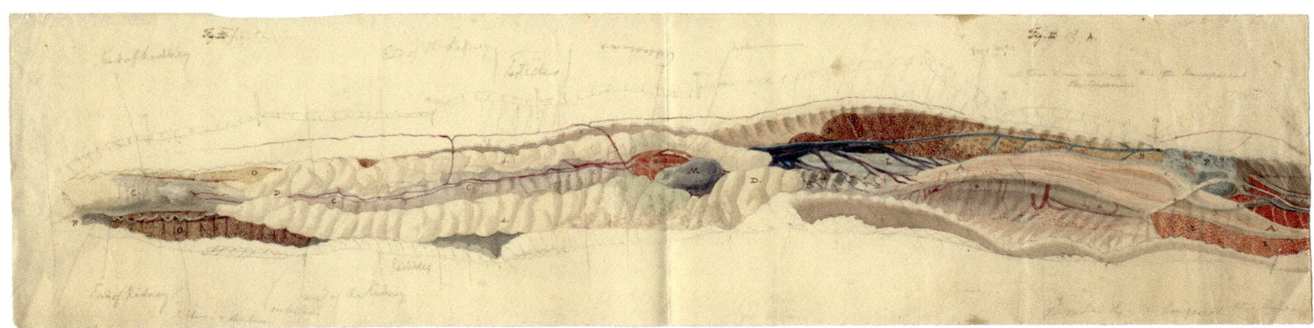

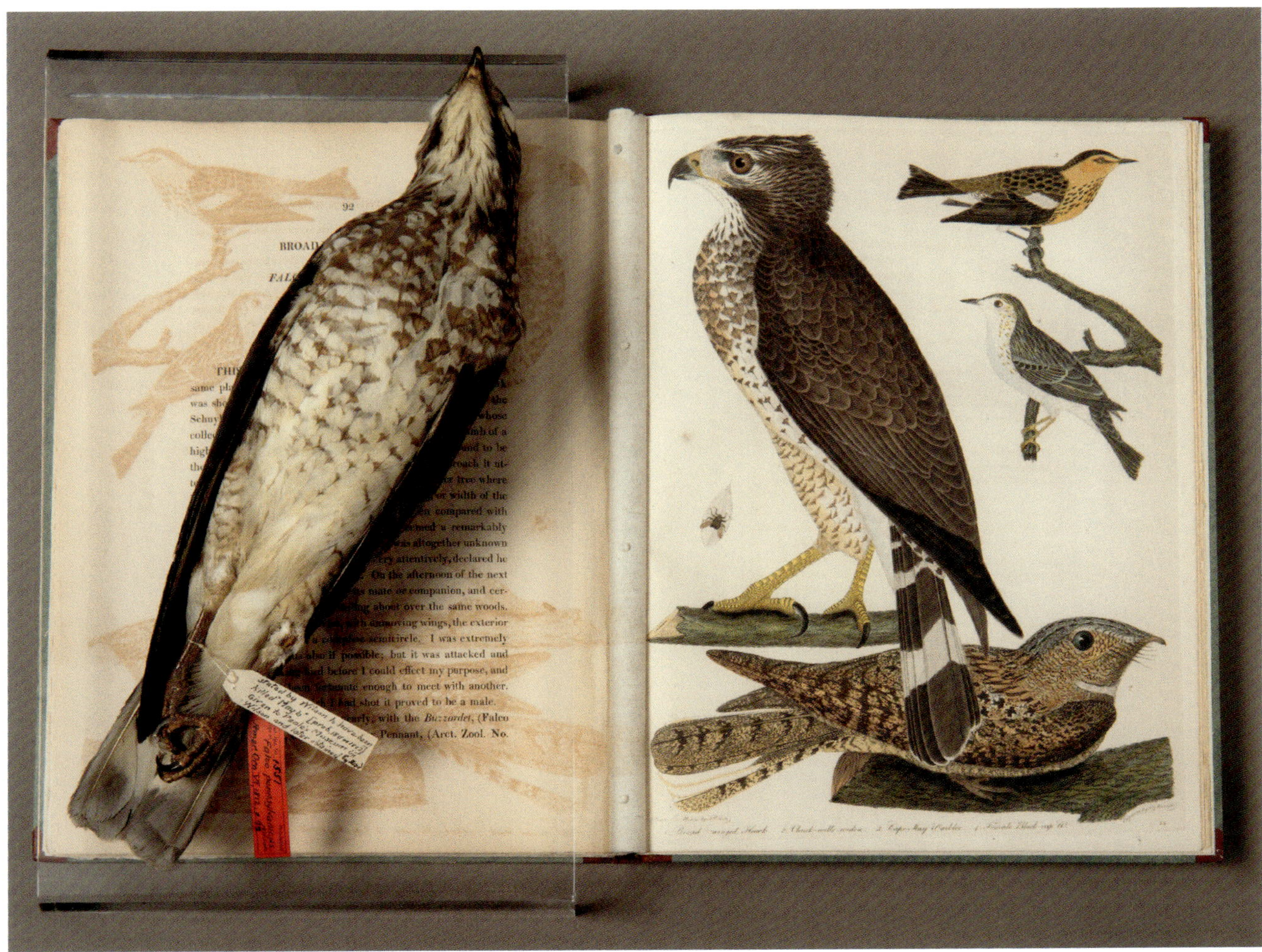

This hawk was collected by Alexander Wilson at Bartram's Garden. Wilson used this specific specimen to describe and name this species, a type of specimen also known as a holotype. It formed the basis for the illustration in *American Ornithology*.

CAT 33
BROAD-WINGED HAWK

Collected at Bartram's Woods,
Schuylkill River, Philadelphia,
PA by Alexander Wilson, 1812
Ornithological specimen
The Academy of Natural
Sciences of Drexel University.

Wilson wrote the first definitive American ornithology. He was instructed in natural history by Bartram and his niece Anne. Wilson's primary goal was to catalog American nature but he also included poetic verse that echoed Bartram's *Travels*.

CAT 34
AMERICAN ORNITHOLOGY,
VOL. 6
Alexander Wilson

Philadelphia, 1812
Hand-colored engraving
APS. Presented by the author
and publisher, 1810–1814.

Sir   Monticello Apr. 7. 05.

I recieved here yesterday your favor of Mar. 18. with the elegant drawings of the new birds you found on your tour to Nia-gara, for which I pray you to accept my thanks. the Jay is quite unknown to me. from my observations while in Europe, on the birds ^& quadrupeds of that quarter, I am of opinion there is not in our continent a single bird or quadruped which is not sufficiently unlike all the members of it's family there to be considered as specifically different. on this general observation I conclude with confi--dence that your Jay is not a European bird.

The first bird on the same sheet I judge to be a Muscicapa from it's bill, as well as from the following circumstance. two or three days before my arrival here a neighbor killed a bird, unknown to him, & never before seen here as far as he could learn. it was brought to me soon after I arrived; but in the dusk of the evening, & so putrid that it could not be approached but with disgust. but I return

mr. wilson

Thomas Jefferson viewed natural history as a means to champion American identity. In this letter, Jefferson thanks Alexander Wilson for drawings of birds and compares them to species known to Georges-Louis Leclerc de Buffon. In his publications on natural history, Jefferson disputed Buffon's claim that American species were weaker than their European counterparts.

CAT 35 (LEFT)
LETTER TO ALEXANDER WILSON
Thomas Jefferson

Monticello, April 7, 1805
Manuscript
APS.

CAT 36 (BELOW)
PARTIAL COPY AFTER ALEXANDER WILSON'S *AMERICAN ORNITHOLOGY*
Thomas Howitt

Lancaster, UK, 1827
Watercolor and ink on paper
APS.

Thomas Howitt was a British amateur ornithologist who made a partial copy of Alexander Wilson's *American Ornithology*. Howitt made changes to Wilson's original by isolating a single bird on each page. He also focused on the scientific descriptions, condensing or cutting Wilson's discussion of bird habits and his personal anecdotes.

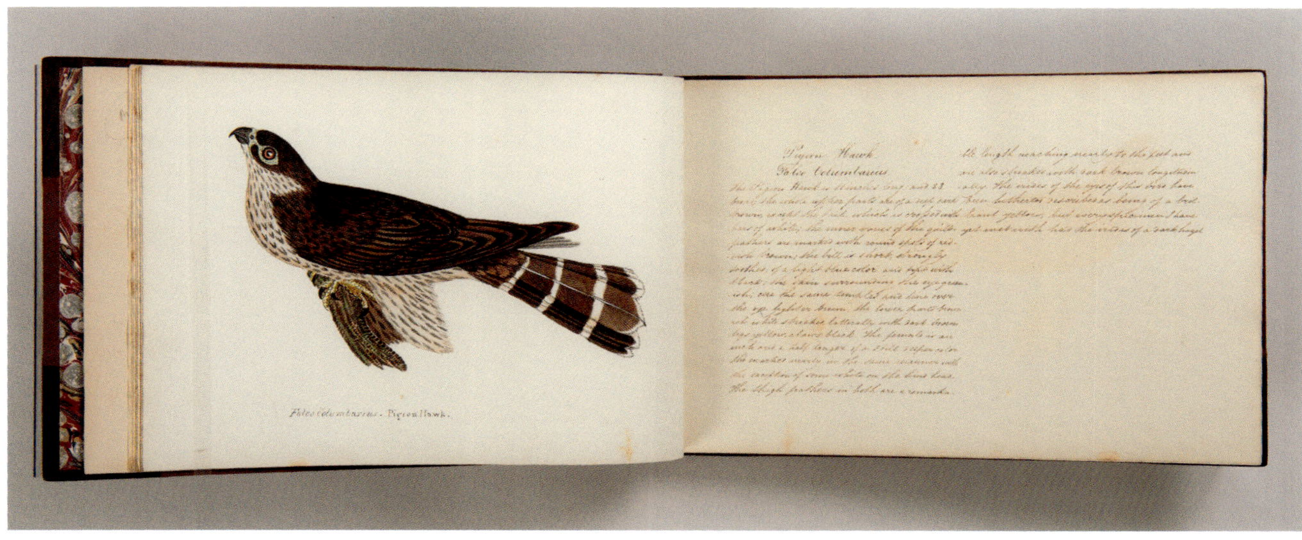

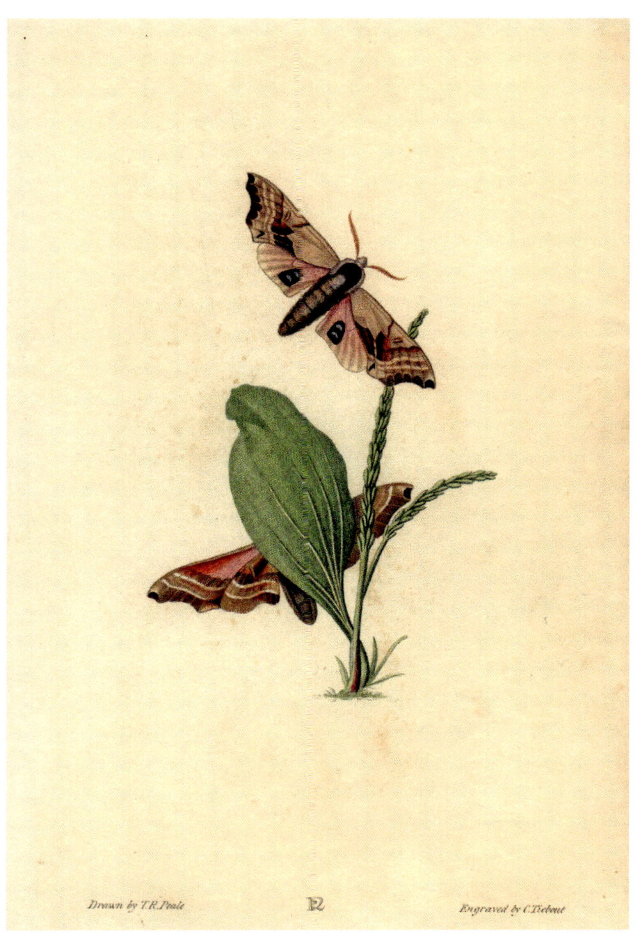

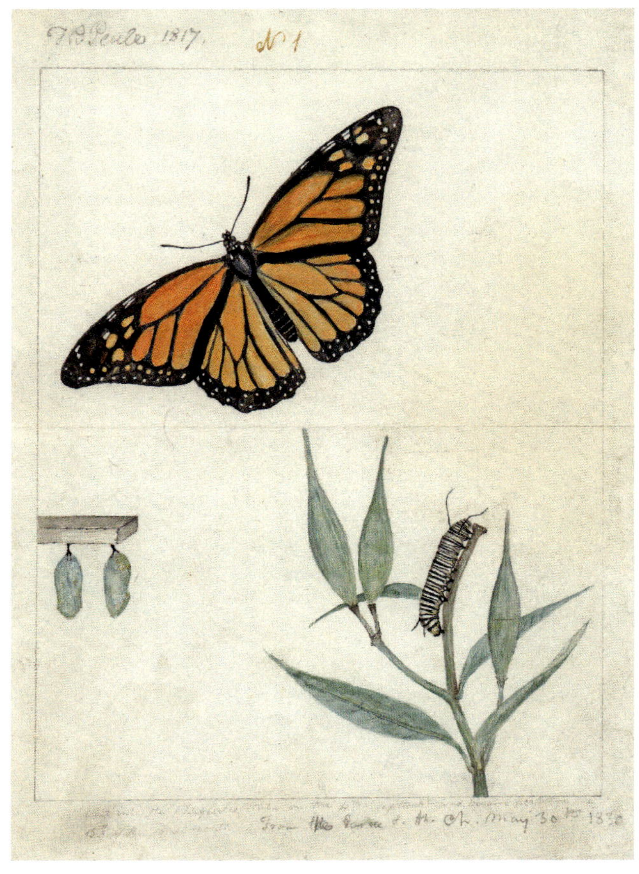

Say was William Bartram's grandnephew. Say's book on insects stuck to rigorous taxonomic descriptions to demonstrate his scientific credentials, as opposed to Bartram's varied approach.

CAT 37
TWIN-SPOTTED SPHINX
C. Tiebout, after Titian
Ramsay Peale

Philadelphia, 1824–1828
Hand-colored engraving
In *American entomology* . . . by
Thomas Say
APS. Presented by the author,
December 1823.

Thomas Say went on field expeditions with members of the Peale family, including Titian Ramsay Peale. At 18, Titian was already observing insects and making detailed drawings of their life cycles, like this one. Thanks to Titian's talent for depicting bugs, Say asked him to illustrate an entomological book.

CAT 38
MONARCH BUTTERFLY
Titian Ramsay Peale

1817
Watercolor and graphite on
paper
APS.

Titian Ramsay Peale

UNFAMILIAR LANDSCAPES

Titian Ramsay Peale (APS, 1833) belonged to a generation of naturalists whose goal was to catalog all native American flora and fauna. His father, Charles Willson Peale (APS, 1786), was a painter by training who founded one of the first American natural history museums. As a youth, Titian was also inspired by William Bartram and his commitment to fieldwork, though Titian embraced taxonomy more wholeheartedly. From 1819 to 1820, Titian went on a government-funded survey between the Mississippi River and the Rockies led by Stephen H. Long (APS, 1823). Titian's work on the expedition focused on describing the species he encountered in these unfamiliar landscapes. He also presented the lands west of the Mississippi as ideal for white settlement in images that are both beautiful and harmonious.

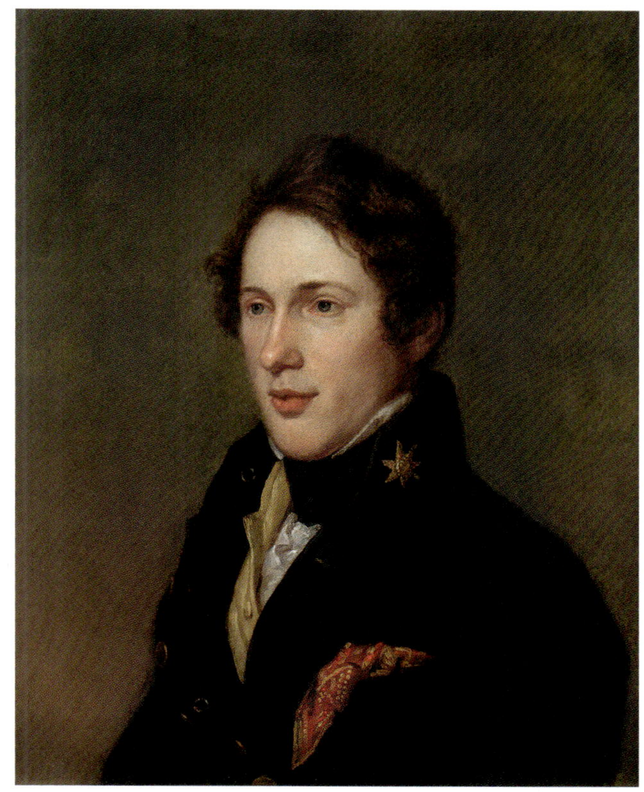

FIG 16 PORTRAIT OF TITIAN RAMSAY PEALE
Charles Willson Peale, 1819, Oil on canvas, Philadelphia Museum of Art: Promised gift of the McNeil Americana Collection, 147-2018-20.

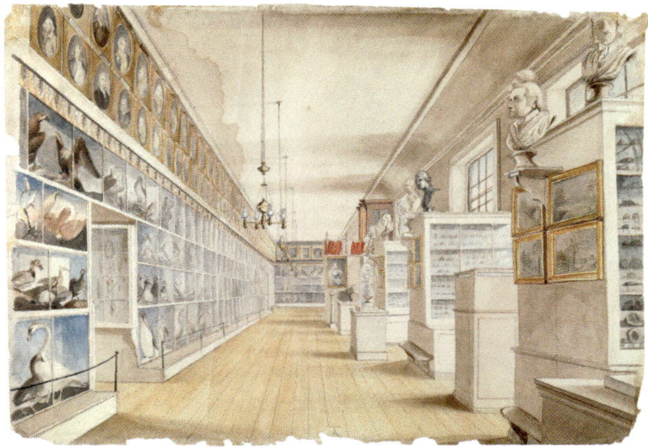

FIG 17 THE LONG ROOM, INTERIOR OF FRONT ROOM IN PEALE'S MUSEUM
Charles Willson Peale; Titian Ramsay Peale, 1822, Watercolor over graphite pencil on paper, Detroit Institute of Arts, Founders Society Purchase, Director's Discretionary Fund, 57.261.

The Peale Museum expanded into the State House in 1802. Charles's bird collection was in the 'Long Room'. It was arranged in neat, gridded rows according to Linnaean taxonomy.

Charles Willson Peale promoted public interest in natural history through his museum, which was housed in the American Philosophical Society's Philosophical Hall from 1794 to 1811. There, visitors could find a range of American animal and mineral specimens arranged in taxonomic order. Each animal was also depicted in a pleasing landscape–remember Peale was trained as a painter–that reflected their native habitat. Charles Willson Peale presented a vision of American nature as harmonious, orderly, and benevolent as the nation expanded across the continent. The museum served as his son Titian's first classroom and he would eventually serve as its curator.

The Peale Museum was the repository for specimens collected on government-run expeditions. This drawing represents two bird specimens from the Lewis and Clark Expedition that were displayed in the museum. Displaying animals from new territories did more than illustrate animal species, it also expressed territorial control.

CAT 39
MOUNTAIN QUAIL AND LEWIS'S WOODPECKER
Charles Willson Peale

Philadelphia, c. 1806
Watercolor over graphite
on paper
APS.

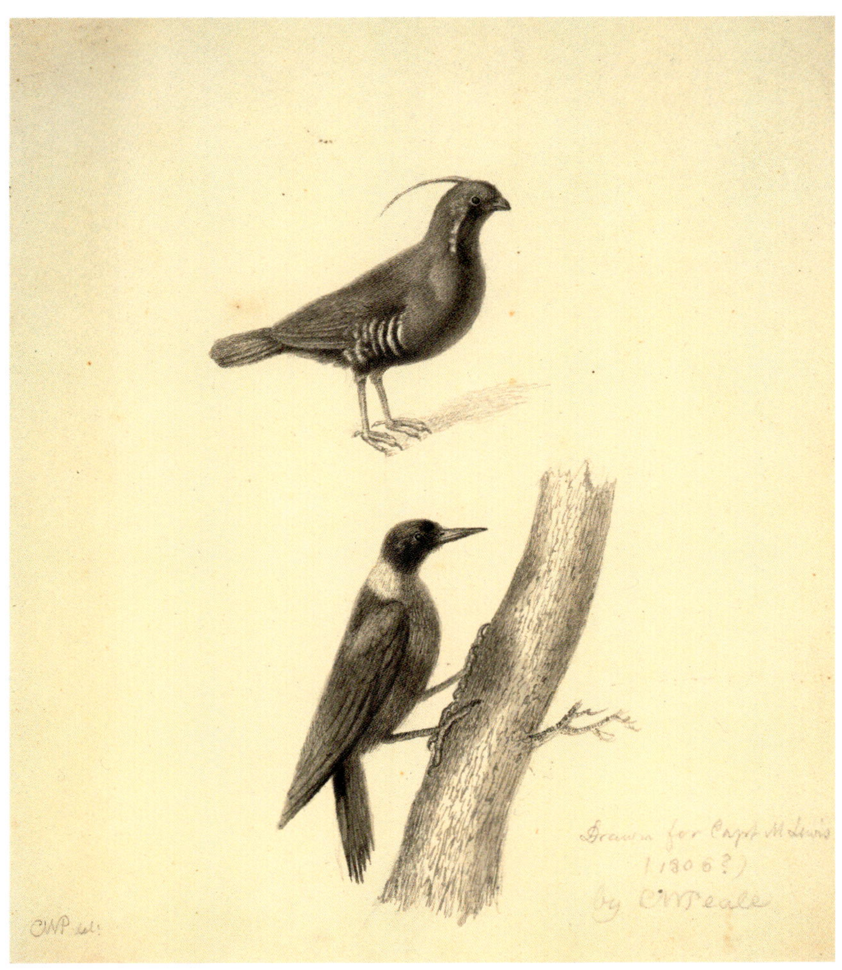

CAT 40
GRIZZLY BEAR
Titian Ramsay Peale

Philadelphia, 1822
Watercolor and graphite
on paper
APS.

This watercolor records two live grizzlies from Zebulon Pike's expedition to the Arkansas and Red Rivers that were sent to Charles Willson Peale. Peale kept them as entertainment for visitors but they were killed when one broke out of its cage.

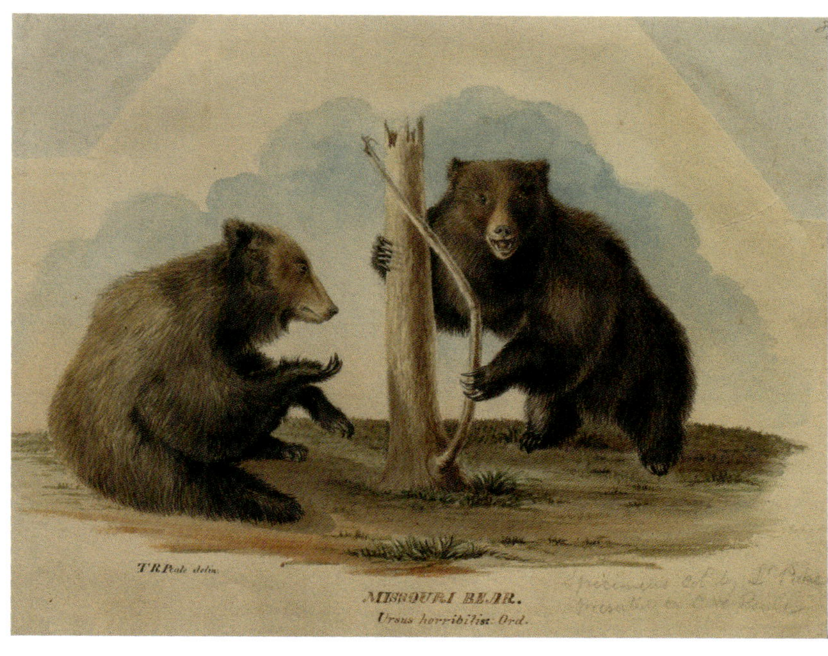

CAT 41
LECTURE ON NATURAL HISTORY
Charles Willson Peale

N.D.
Manuscript
APS.

Charles Willson Peale delivered public lectures that connected natural history to ethics. Here he uses ecological ideas to deliver a moral message, that each living thing deserves respect because it performs an essential function. At the same time Peale believed hierarchical distinctions based on race, gender, and class were also 'natural.'

The Lewis and Clark Expedition (1804–1806) was the first American expedition west of the Mississippi, but the Long Expedition (1819–1820) was the first to bring trained naturalists and artists. Titian Ramsay Peale's finished watercolors are idealized scientific illustrations of the animal's appearance, its habitat, something of its behavior, and its diet. These images also clearly communicate the scientific information required by the government: what animals live there, the quality of the land, its plants, and the availability of water. Peale's harmonious presentation of nature made it seem open to settlement. A few images, however, depict a harsher side of the landscape.

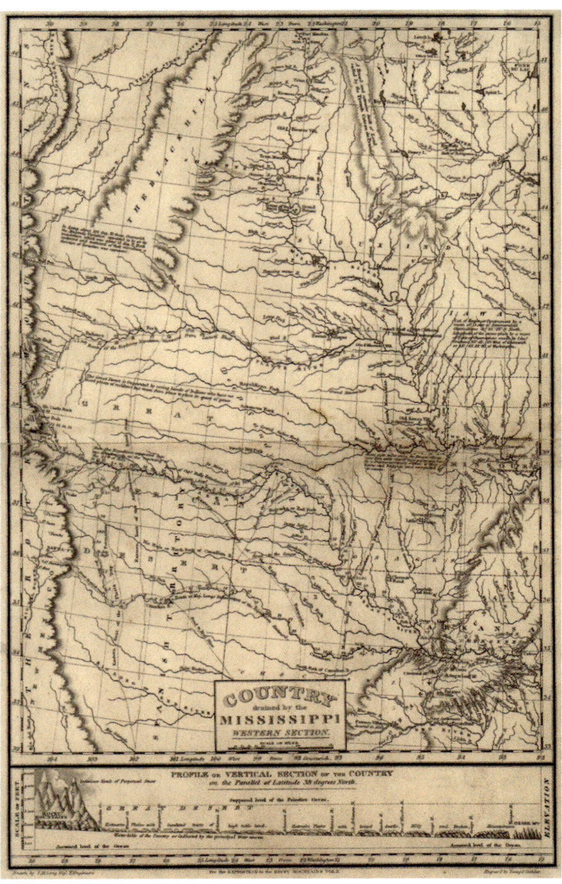

FIG 18 COUNTRY DRAINED BY THE MISSISSIPPI, WESTERN SECTION
Stephen H. Long, 1822, Engraving, Library of Congress, Geography and Map Division.

Stephen Long's map shows the various routes taken by the expedition's members between the Missouri River and the Rockies, a region Long labeled 'The Great American Desert.'

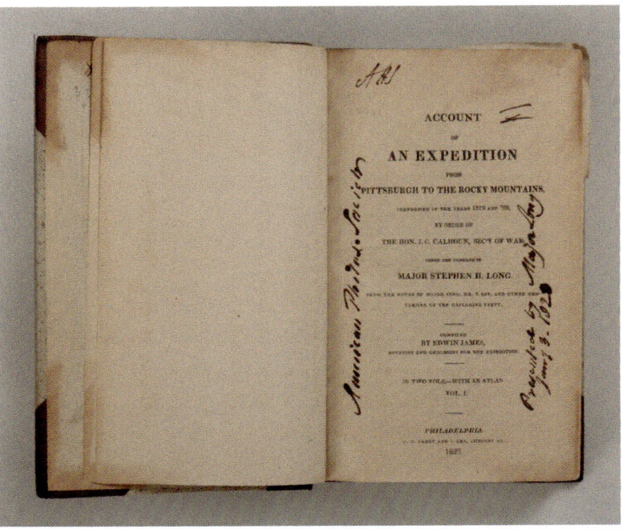

The Long Expedition spent the winter near Council Bluffs (modern Nebraska) and set off in June for the Rocky Mountains. The journey was difficult. The official account described the whole region from the Missouri to the Rockies as a 'Great American Desert' and recommended against its settlement.

CAT 42
ACCOUNT OF AN EXPEDITION FROM PITTSBURGH TO THE ROCKY MOUNTAINS
Edwin James

Philadelphia, 1822
Bound volume
APS. Gift of Major
Stephen H. Long, 1823.

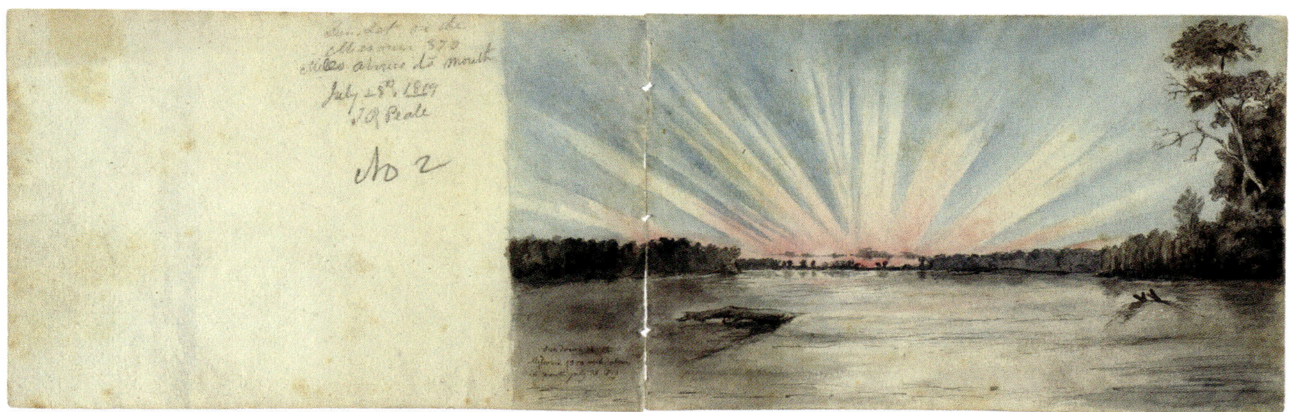

Titian Ramsay Peale made this watercolor early in the Long Expedition, after they left St. Louis and headed up the Missouri River. The lure of the glowing sun, setting in the west, illustrates his belief that westward expansion was central to the United States' future.

CAT 43 (ABOVE)
SUNSET ON THE MISSOURI
Titian Ramsay Peale

July 28, 1819
Watercolor, graphite,
and ink on paper
APS.

Titian Ramsay Peale painted the expedition's winter headquarters on the Missouri River, christened 'Engineer Cantonment,' just south of Council Bluffs. The expedition's steamboat is shown flying an American flag and billowing smoke in a wintry landscape, a symbol of technological might.

CAT 44 (BELOW)
ENGINEER CANTONMENT
Titian Ramsay Peale

Engineer Cantonment,
February 1820
Watercolor, graphite,
and ink on paper
APS.

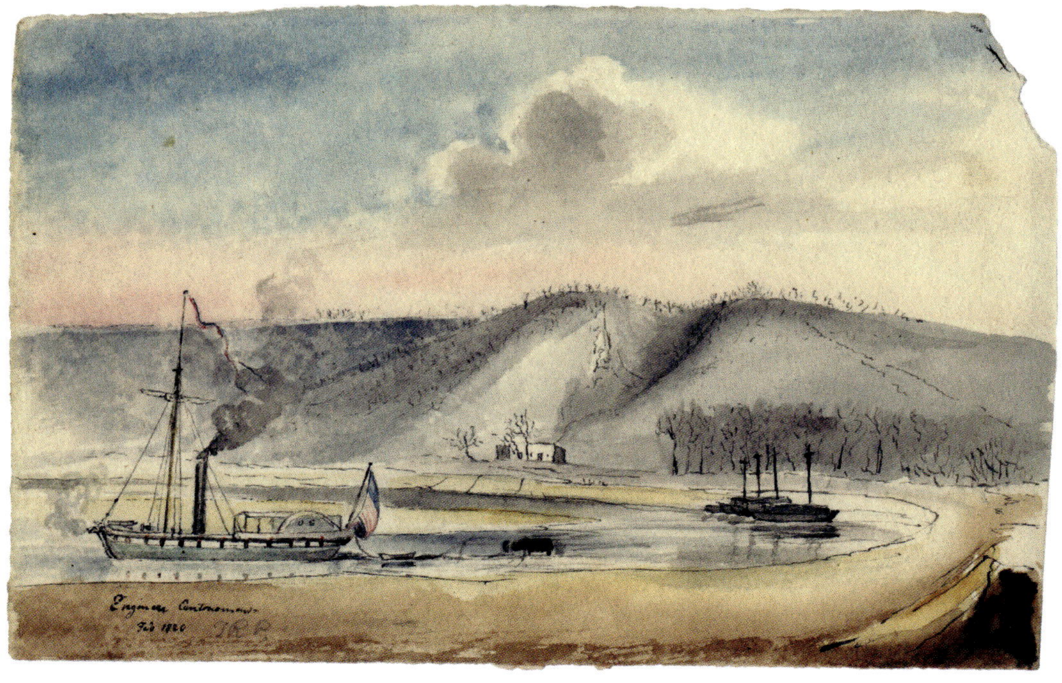

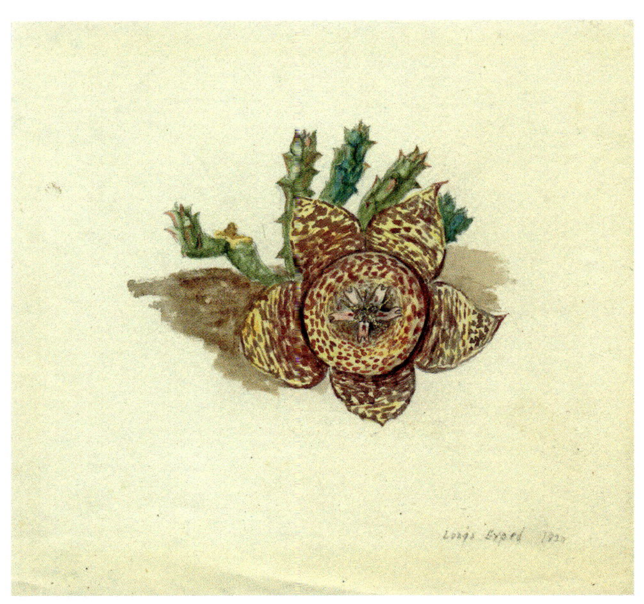

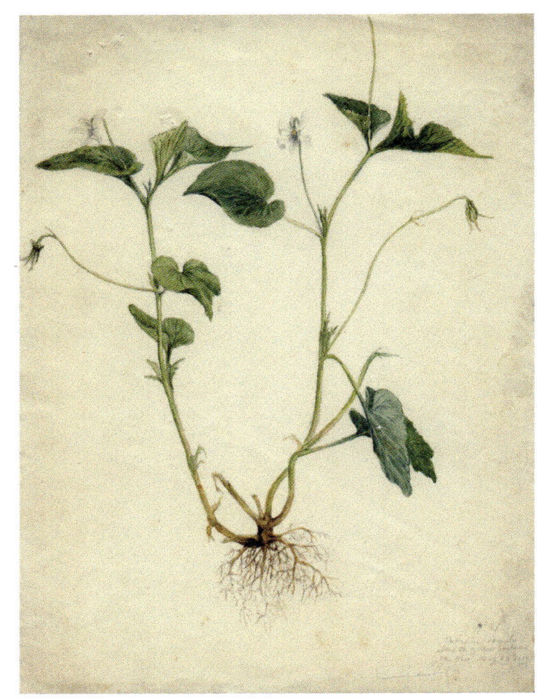

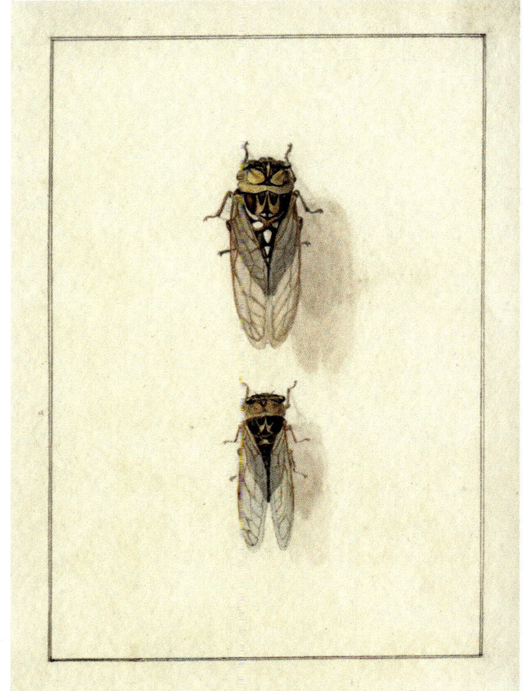

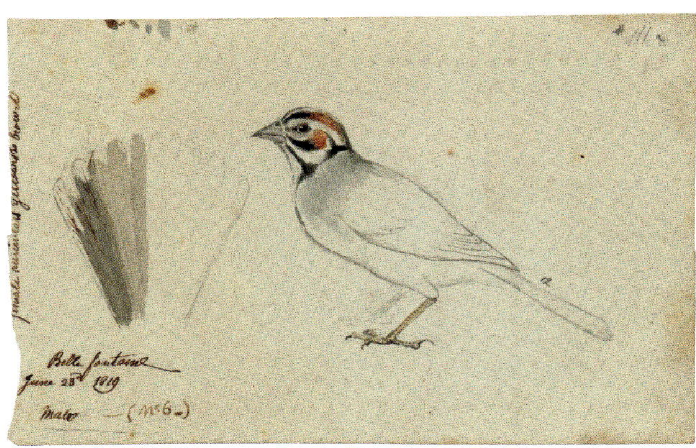

Titian Ramsay Peale's finished Long Expedition watercolors, like the Blacktail Deer and Muskrats, bring together his detailed botanical and zoological studies, as well as his observations of animal behavior. In many watercolors, he presents an aesthetically pleasing portrait of the landscape, with abundant water, lush plant life, and rosy sunsets. However, the Long Expedition account presented much of the land between the Missouri River and the Rockies as harsh and unpromising for white settlement. Peale evokes this sentiment in some of his images, where storm clouds gather, or animals fight for scraps of food.

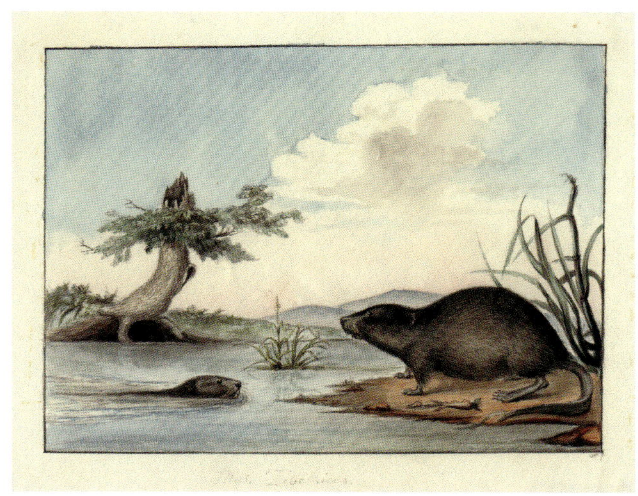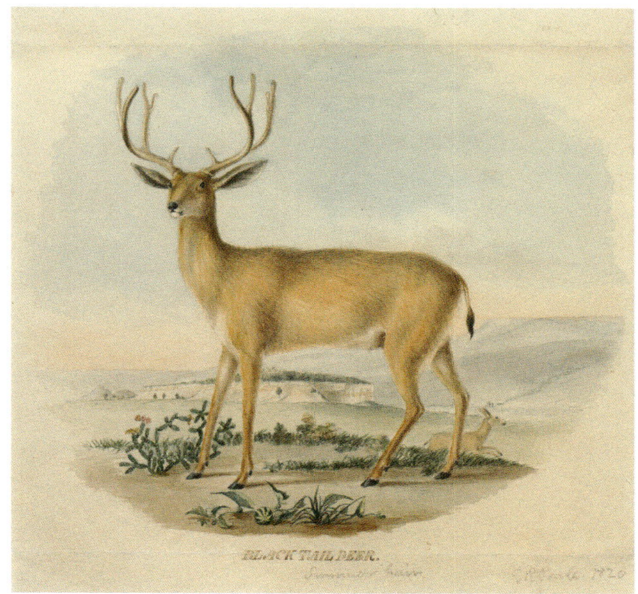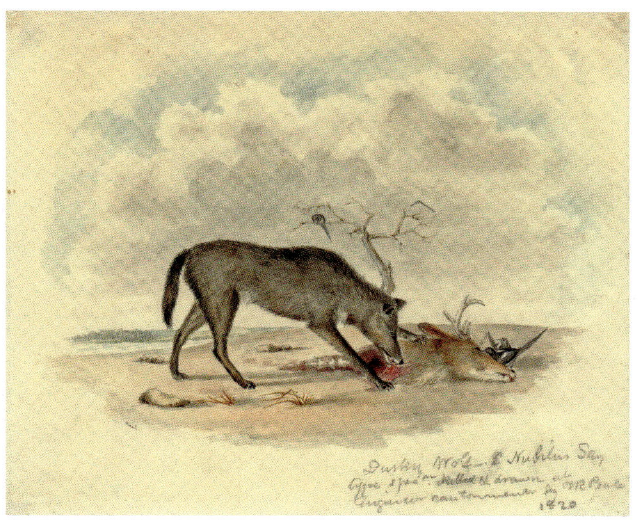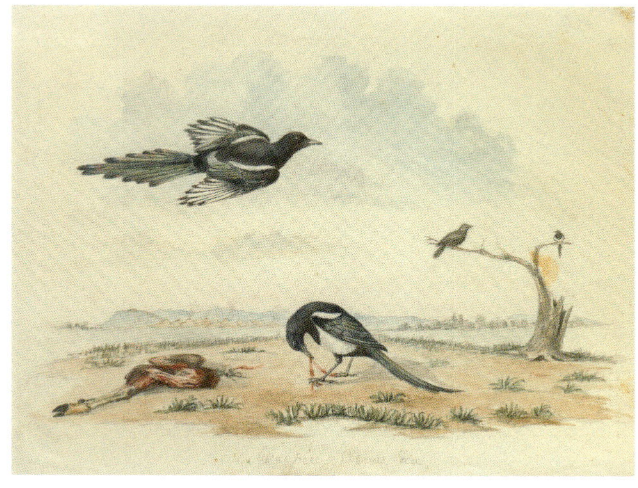

CAT 45
CACTUS FLOWER
Titian Ramsay Peale

1820
Watercolor and graphite
on paper
APS.

CAT 46
WHITE VIOLET
Titian Ramsay Peale

Indiana, May 25, 1819
Watercolor and graphite
on paper
APS.

CAT 47
MUSKRATS
Titian Ramsay Peale

1819–1820
Watercolor and graphite
on paper
APS.

CAT 48
**BLACKTAIL DEER,
SUMMER HAIR**
Titian Ramsay Peale

1822
Watercolor, graphite,
and ink on paper
APS.

CAT 49
CICADAS
Titian Ramsay Peale

1819–1820
Watercolor on paper
APS.

CAT 50
SPARROW
Titian Ramsay Peale

June 23, 1819
Watercolor, graphite,
and ink on paper
APS.

CAT 51
**DUSKY WOLF DEVOURING
DEER HEAD**
Titian Ramsay Peale

1821–1822
Watercolor and graphite
on paper
APS.

CAT 52
MAGPIE
Titian Ramsay Peale

June 29, 1820
Watercolor and
graphite on paper
APS.

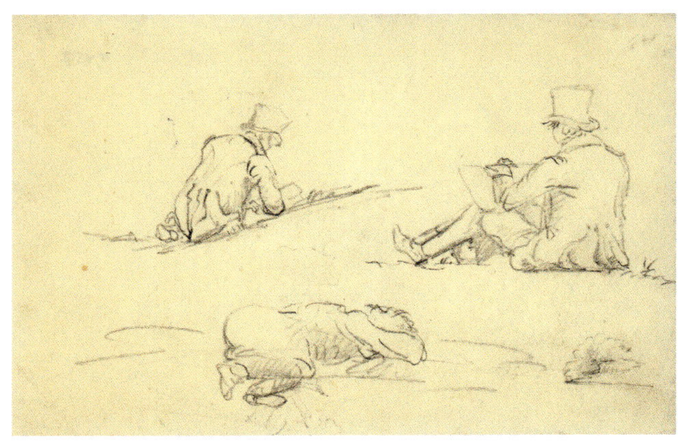

HUNTING ANIMALS, MAKING SCIENCE

In order to study animals, naturalists often killed them. One of Titian Ramsay Peale's main roles during the Long Expedition was hunting for both food and scientific specimens. Sometimes, a single animal served both functions. The violence behind the science usually goes unrecorded, but Peale produced a series of images that record trapped or dead animals. In these images, he seems to present his art as something that can restore life after death. The dead scarlet tanager is already restored to the brilliant color it had in life, and taxidermy will complete the resurrection. But these images also highlight the relationship between scientific knowledge and violent control.

This sketch may be a self-portrait of Peale while drawing. It confirms his strong identification with the role of the draftsman. He also shows himself drawing in various awkward positions suggesting the physical discomforts of work in the field.

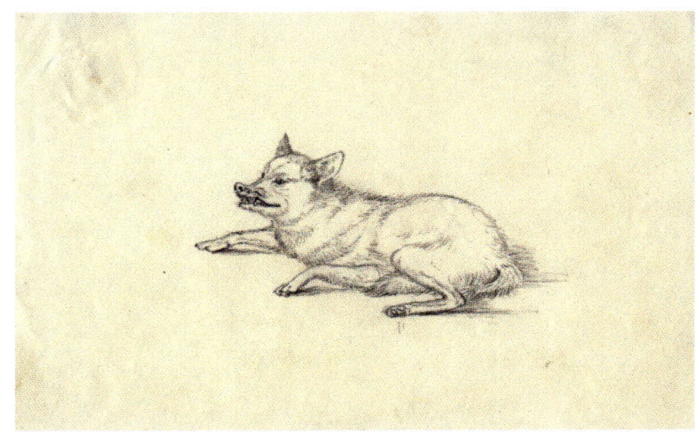

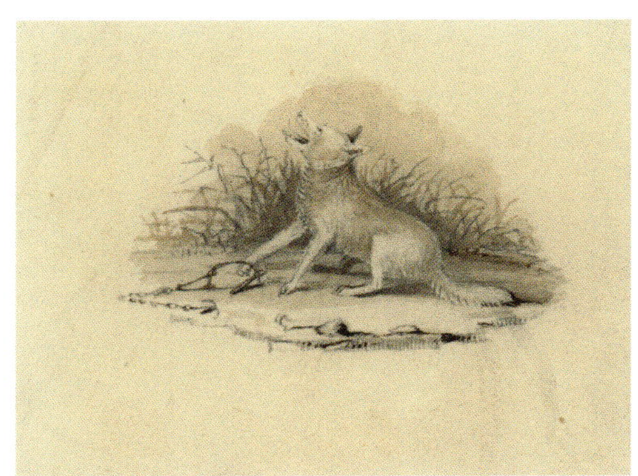

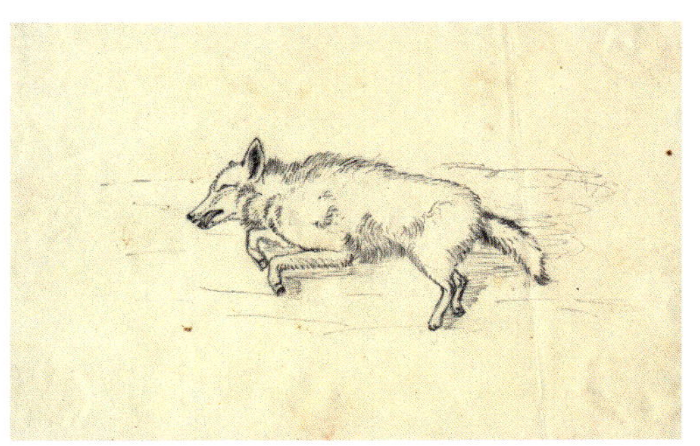

CAT 53 (ABOVE)
SELF-PORTRAIT (?)
SKETCHING
Titian Ramsay Peale
N.D.
Graphite on paper
APS.

CAT 54 (RIGHT, TOP)
GREAT PLAINS WOLF
SNARLING
Titian Ramsay Peale
1819–1820
Graphite and ink on paper
APS.

CAT 55 (RIGHT, CENTER)
GREAT PLAINS WOLF
IN DISTRESS
Titian Ramsay Peale
1819–1820
Graphite and ink wash
on paper
APS.

CAT 56 (RIGHT, BOTTOM)
GREAT PLAINS WOLF
DEAD
Titian Ramsay Peale
1819–1820
Graphite and ink on paper
APS.

It was unusual for naturalists to represent the violent aspects of natural history, but Peale did just that in this series of three images of a Great Plains Wolf. A first image shows the wolf snarling and hostile, a second trapped and emotionally distressed, and the third finally dead after Peale himself had shot it. This series invites reflection about the impact of scientific discovery on its subjects, as the creation of new knowledge came at a cost.

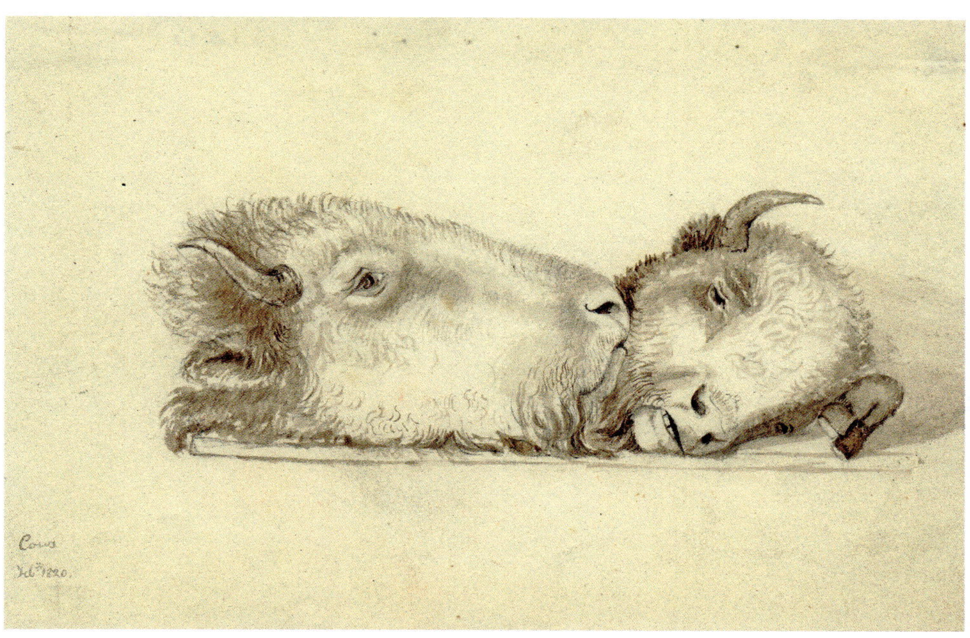

CAT 57
BISON HEADS
Titian Ramsay Peale

February 1820
Graphite and ink wash
on paper
APS.

In this unsettling image, the bison on the right is dead, while the head on the left stares out as if still alive. Both are displayed on a shelf that recalls a museum exhibit, anticipating their future as taxidermied displays. Peale's image suggests that art and taxidermy were ways of 'resurrecting' animals.

CAT 58
SCARLET TANAGER
Titian Ramsay Peale

Engineer Cantonment, 1820
Watercolor, graphite,
and ink on paper
APS.

A lifeless scarlet tanager lies on a sheet of drawing paper. Here Peale highlights art as a key way that the artist-naturalist shared knowledge about nature. Images could communicate information about the living animal (like vibrant color) that faded specimens could not.

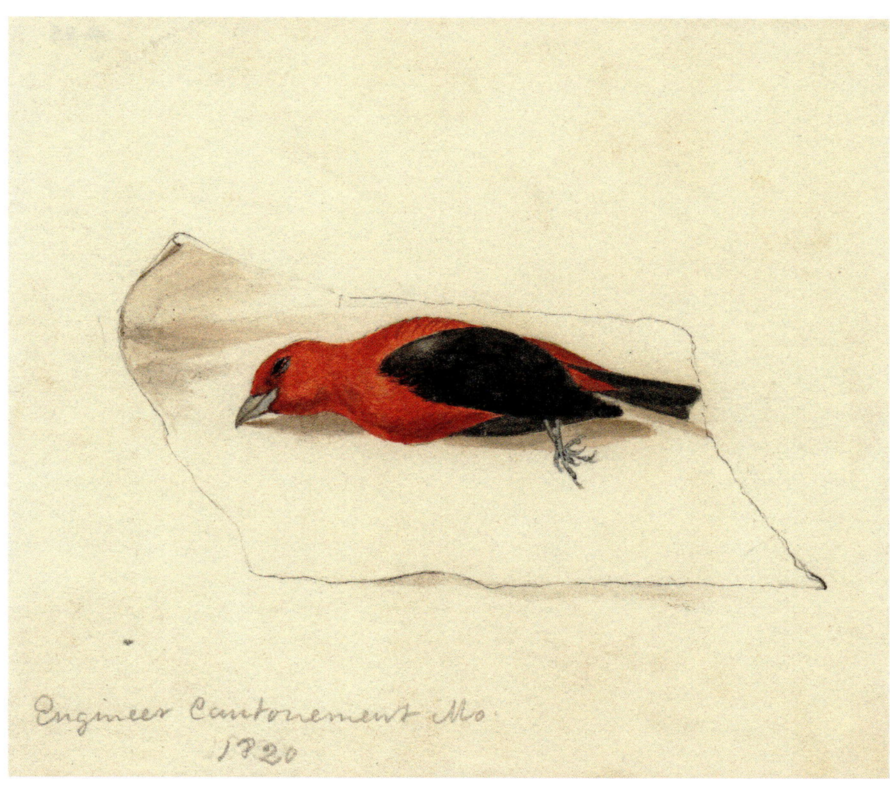

Titian Ramsay Peale sought to organize his experiences in an unfamiliar natural world through art. His initial sketches record raw information gained in the field, including both an animal's appearance and its behavior. More finished studies were used to refine the animal's appearance. Rather than offering a portrait of an individual, Peale wanted to record the characteristic or ideal qualities of a whole species. Finally, he brought together his concrete field observations with the idealized specimen image to communicate an entire species' appearance and behavior. As he moved from the chaotic initial sketch to an orderly finished product, he asserted control over unwieldy nature.

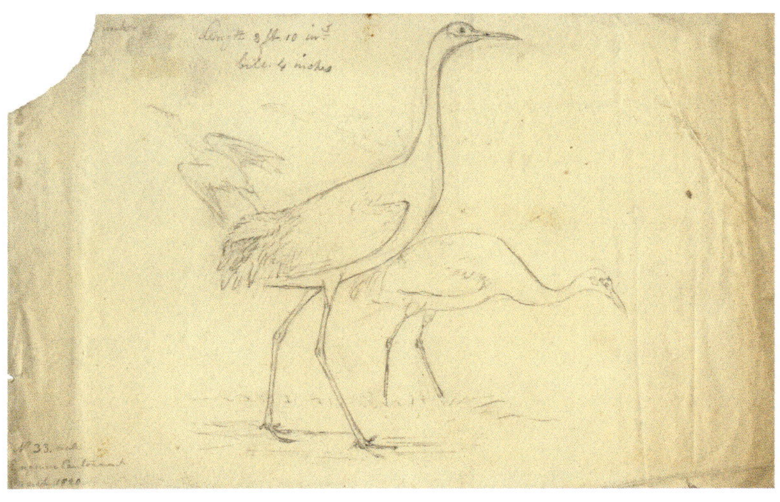

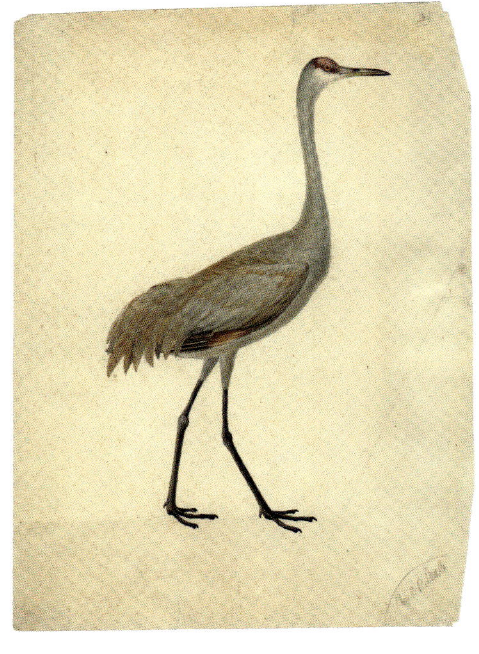

Peale sketched several rough drafts before creating his final images. The initial sketch records first impressions of Sandhill Cranes in their wetlands habitat along the Platte River. Peale depicts the birds' appearance, food, social behavior, and flight. In the watercolor sketch of the single crane, he perfects the lines of its form and accentuates its feathers. The image is no longer a record of an individual bird, but an ideal representation of the species. In the final watercolor, he combines the idealized portrait and observations from the living bird. He also rearranges the whole so that the finished watercolor presents nature as orderly.

CAT 59
SANDHILL CRANES
Titian Ramsay Peale

Engineer Cantonment,
March 1820
Graphite on paper
APS.

CAT 60
SANDHILL CRANE
Titian Ramsay Peale

1820
Watercolor and
graphite on paper
APS.

NATURAL HISTORY ILLUSTRATION:

Naturalists created many types of images. Specimen images separated organisms from their natural contexts, other images situated them in habitats, while still others depicted whole ecosystems.

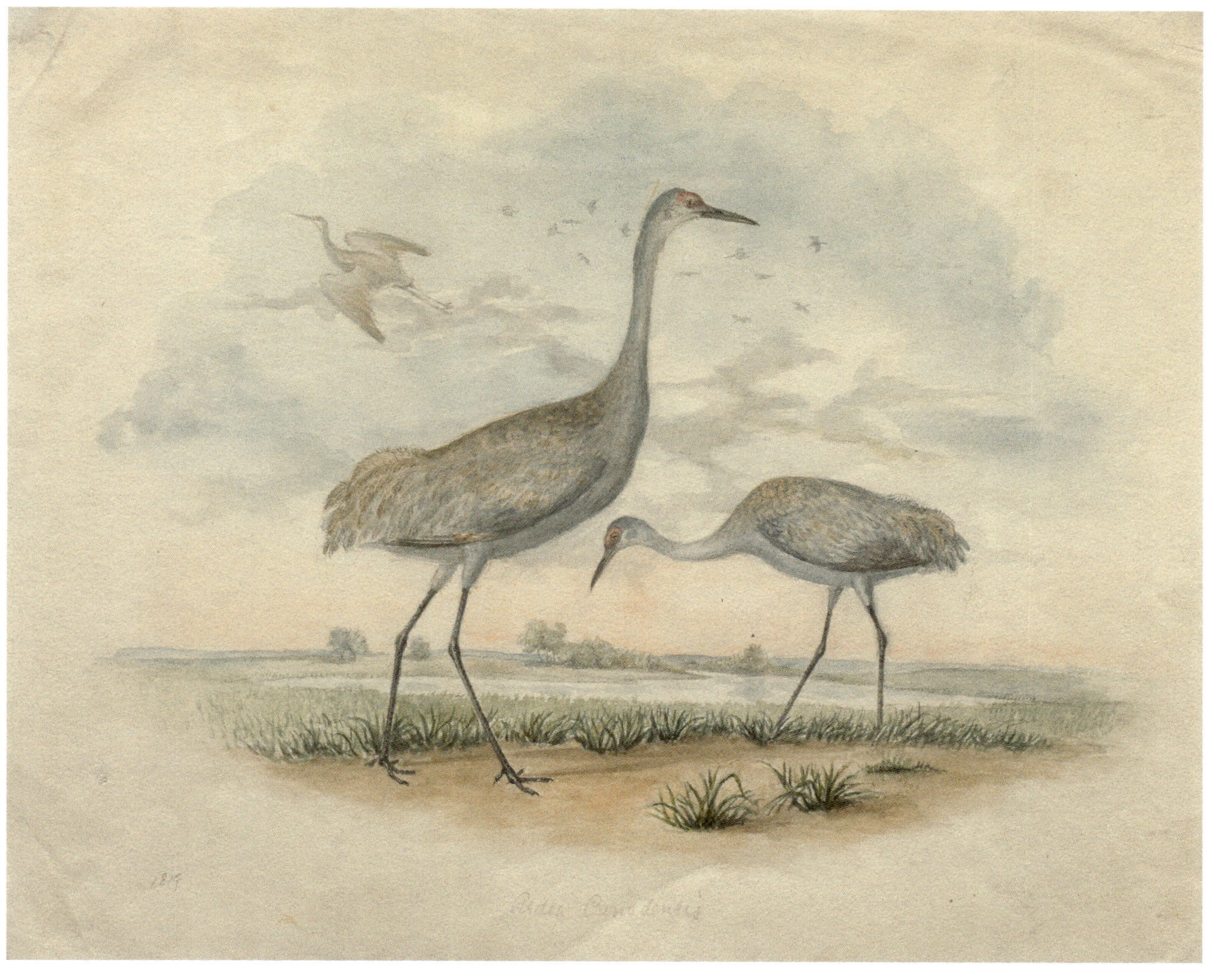

CAT 61
SANDHILL CRANES
Titian Ramsay Peale

1820
Watercolor and
graphite on paper
APS.

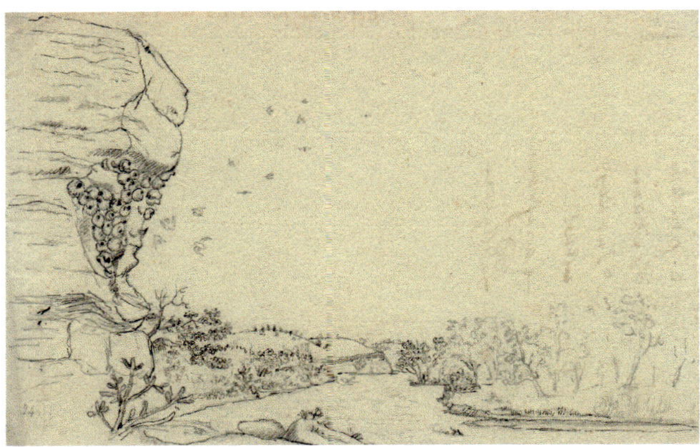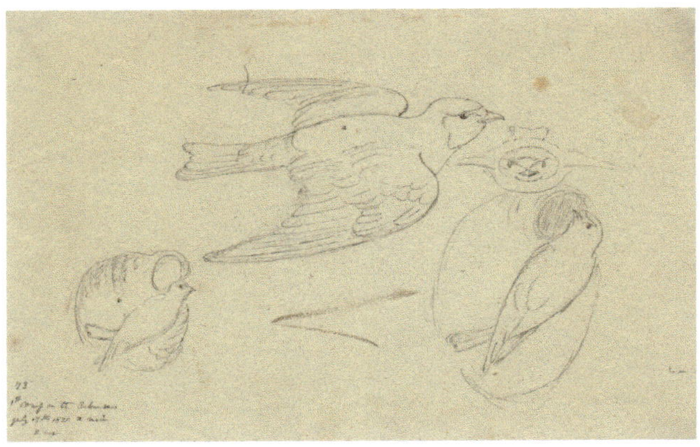

This series documents Peale's process from sketch to published engraving. Peale first made two sketches of cliff swallows in flight and the forbidding cliffs on which the swallow colonies nest and raise their young. He subsequently combined the image of the swallow in flight and its cliff habitat—now even more forbidding—in a second study, this one for a plate in Charles Lucian Bonaparte's *American Ornithology*. Finally, Peale's study was engraved by Bonaparte's printer Alexander Lawson. Lawson made still more changes, modifying the shape of the cliff so that it appears less ominous. From sketch to print, the wilderness becomes ever more domesticated.

CAT 62 (ABOVE, LEFT)
NESTS OF CLIFF SWALLOWS
Titian Ramsay Peale

July 17, 1820
Ink on paper
APS.

CAT 63 (ABOVE, RIGHT)
CLIFF SWALLOWS
Titian Ramsay Peale

July 17, 1820
Ink on paper
APS.

CAT 64 (BELOW, LEFT)
FULVOUS OR CLIFF SWALLOW, BURROWING OWL, STUDY FOR *AMERICAN ORNITHOLOGY*
Titian Ramsay Peale

c. 1825
Graphite and ink wash on paper
Academy of Natural Sciences of Drexel University, Library and Archives.

CAT 65 (BELOW, RIGHT)
FULVOUS OR CLIFF SWALLOW, BURROWING OWL
Alexander Lawson, after Titian Ramsay Peale

Philadelphia, 1825
Hand-colored engraving
In *American Ornithology*, vol. 1
by Charles Lucian Bonaparte
APS.

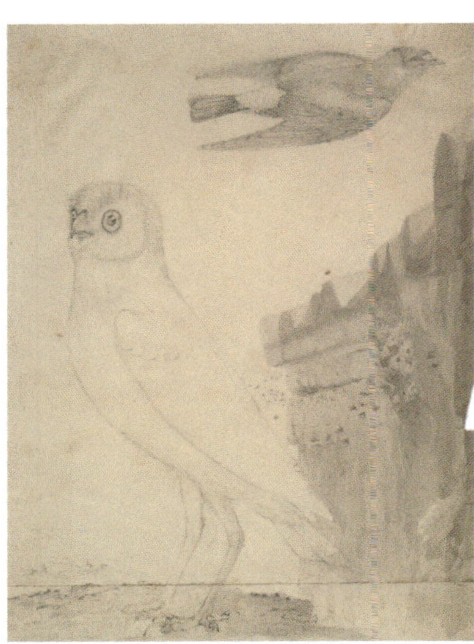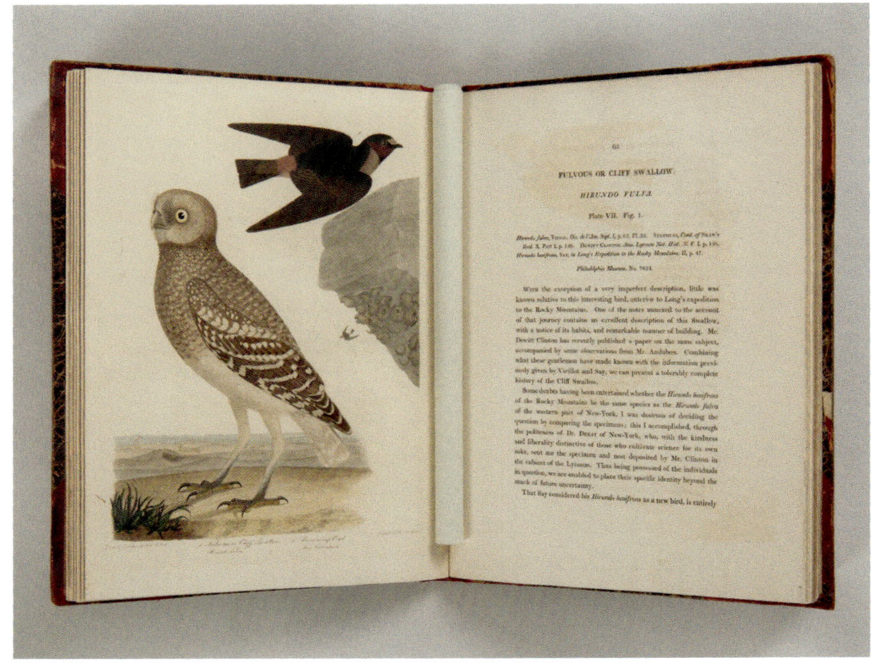

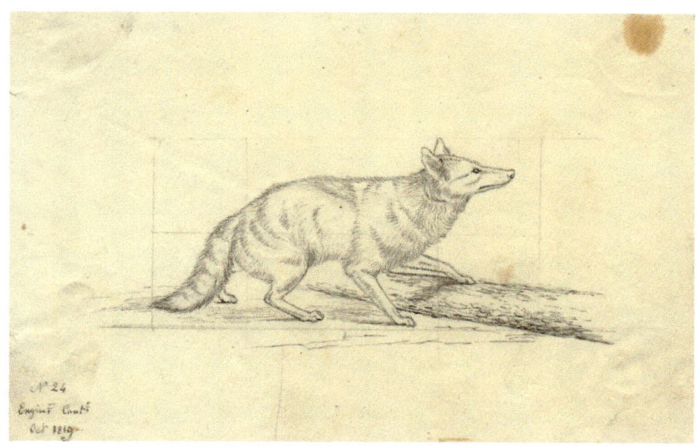

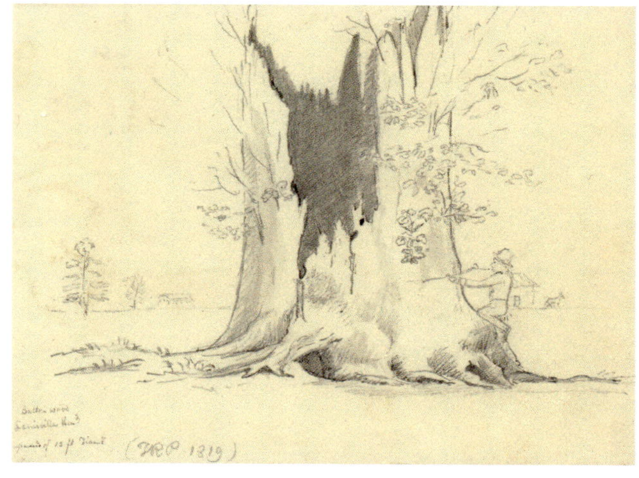

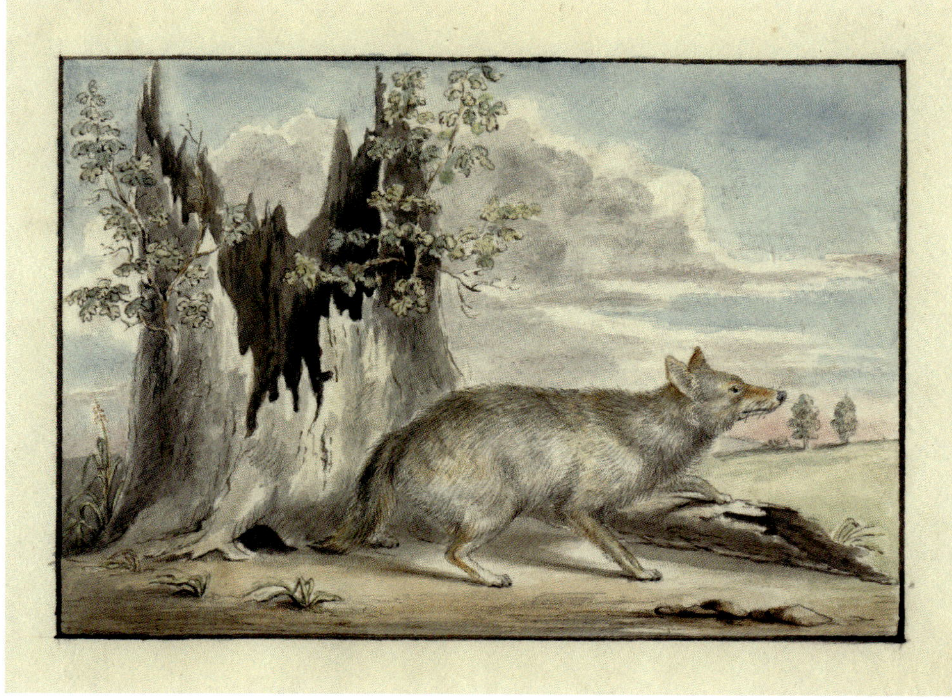

CAT 66
FOX
Titian Ramsay Peale

Engineer Cantonment,
October 1819
Graphite and ink on paper
APS.

CAT 67
**BUTTON WOOD (TREE),
LOUISVILLE KENTUCKY**
Titian Ramsay Peale

Louisville, 1819
Graphite and ink on paper
APS.

CAT 68
FOX
Titian Ramsay Peale

1819–1820
Watercolor, graphite,
and ink on paper
APS.

Peale's images generally focus on delivering accurate scientific information, but in this series, he took more license. In his finished watercolor, Peale combined a sketch of a fox he saw near the expedition's winter headquarters at Engineer Cantonment, with a sketch of an enormous buttonwood tree that he saw thousands of miles away. He also shrunk the tree stump. This unusual working process suggests Peale wanted to convey a larger message. The fox in his watercolor watches warily for a predator, while the green shoots springing from the dead trunk invite meditation on cycles of life and death in nature.

John James Audubon

THE 'UNIVERSE' OF AMERICA

John James Audubon (APS, 1831) was born in the French colony of Saint-Domingue (modern Haiti). He came to the United States in 1803, where he cultivated an interest in ornithology. Audubon's *The Birds of America* (1827–1838) is best known for its dramatic representations of American birdlife. Audubon introduced multi-figure compositions that captured birds' environments and complex social lives. These vivid images of American wildlife intensified the identification of the United States with its natural splendors. But these images also laid intellectual claim to the land as Native Americans faced increased dispossession in the 1820s and 1830s. Further, they romanticized landscapes that supported plantation slavery. Audubon supported these policies both as an enslaver and as a participant in scientific racism.

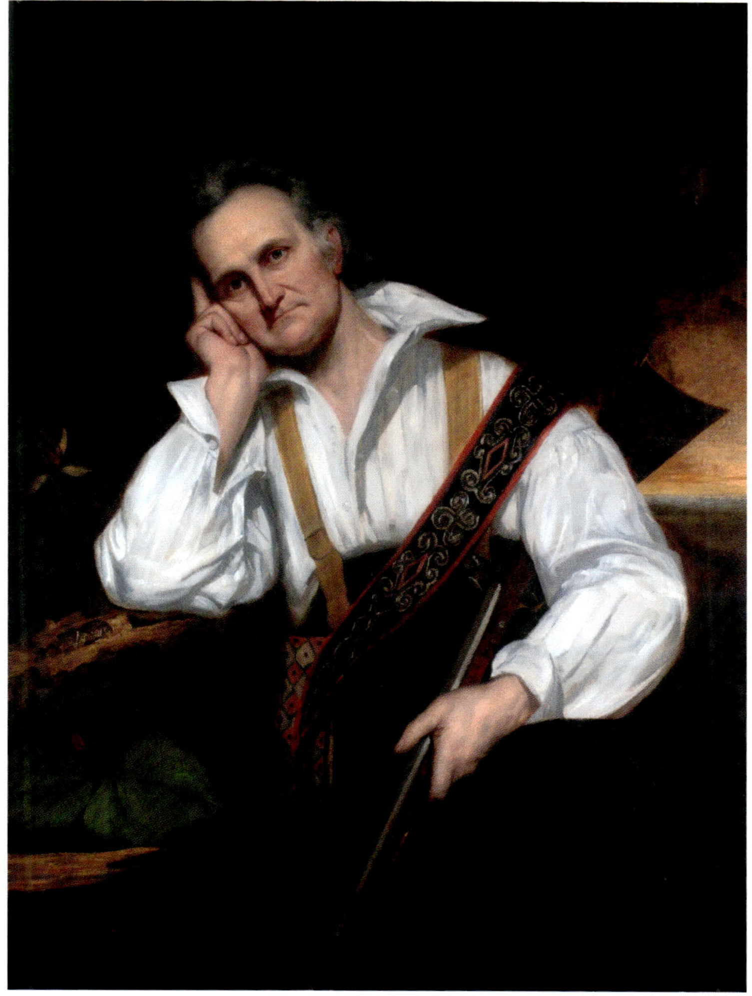

FIG 19 PORTRAIT OF JOHN JAMES AUDUBON
George P.A. Healy, c. 1838, Oil, Museum of Science, Boston.

John James Audubon was an outsider who struggled to gain acceptance from Philadelphia's naturalist community. In *Birds*, Audubon departed from the strict taxonomic approach advocated by professional naturalists. Instead, he staked his reputation on a commitment to unsurpassed fieldwork. To capture the American wildlife he observed firsthand, he showed active birds interacting with one another and their environments. These multi-figure compositions combined designs from contemporary painting and traditional scientific illustration. Audubon also infused his illustrations with drama to express the emotional life of animals. Audubon claimed these designs were totally original, but he built on clear precedents in the work of earlier naturalists.

CAT 69
THE BIRDS OF AMERICA,
VOL. 1
John James Audubon

A (OPPOSITE)
MOCKING BIRD
Plate XXI

B (RIGHT)
COLUMBIA JAY
Plate XCVI

1827–1838
Hand-colored etching, aquatint, and engraving
APS. Subscribed for by divers members and presented to the Society, Oct. 7, 1831.

Audubon identified 25 bird species new to scientists of European descent. He also made new ecological insights based on his close observations of birds in the wild. However, he was not able to convince printers in Philadelphia to publish his unorthodox book, so it was printed in Edinburgh and London.

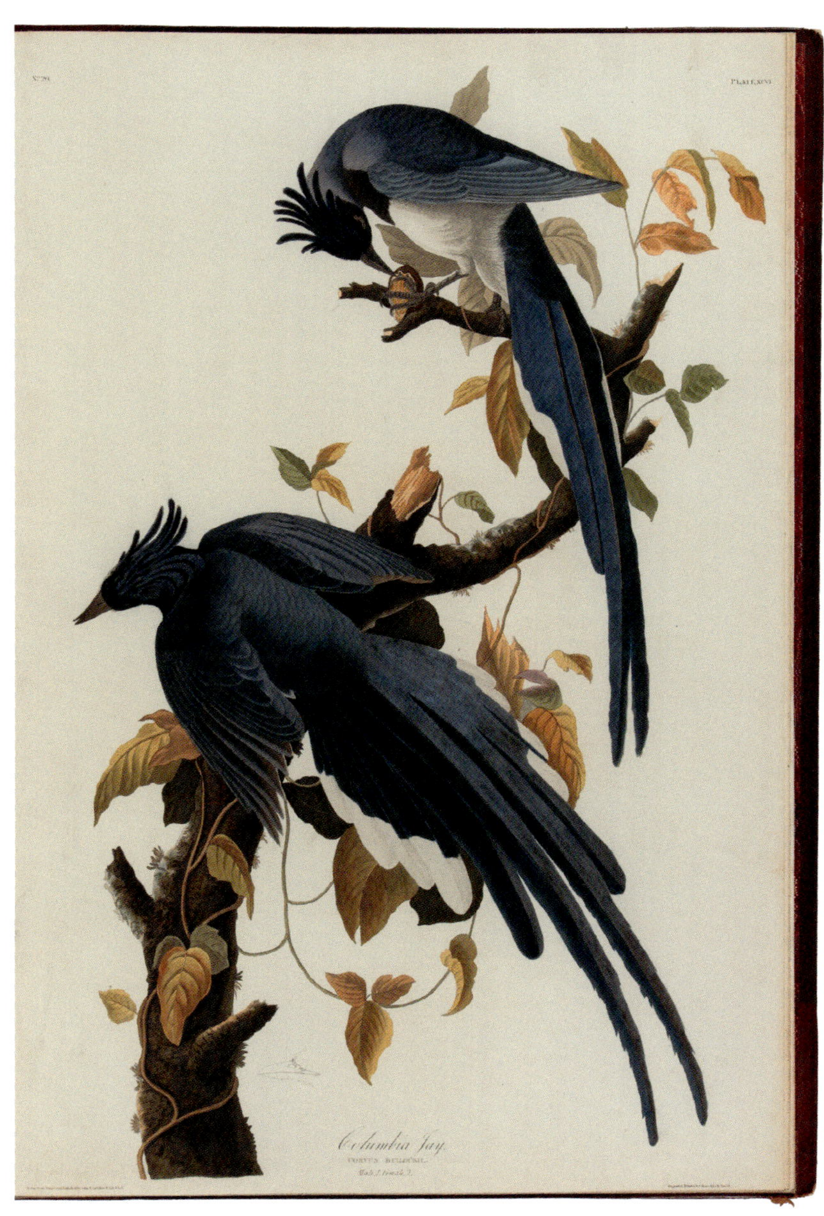

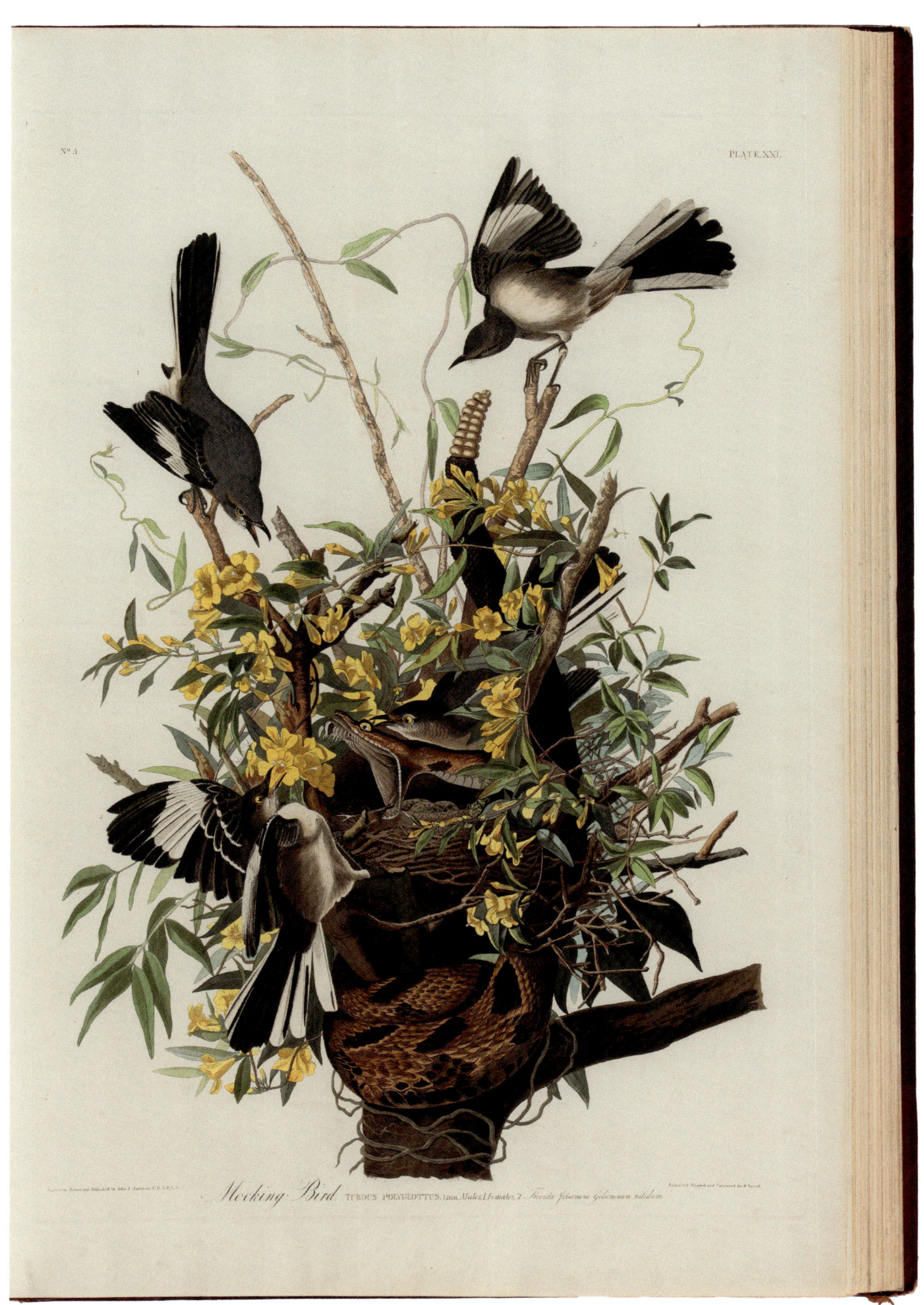

Mocking Bird. TURDUS POLYGLOTTUS, Linn. Males 1. Females, 2. Florida Jessamine Gelseminum nitidum

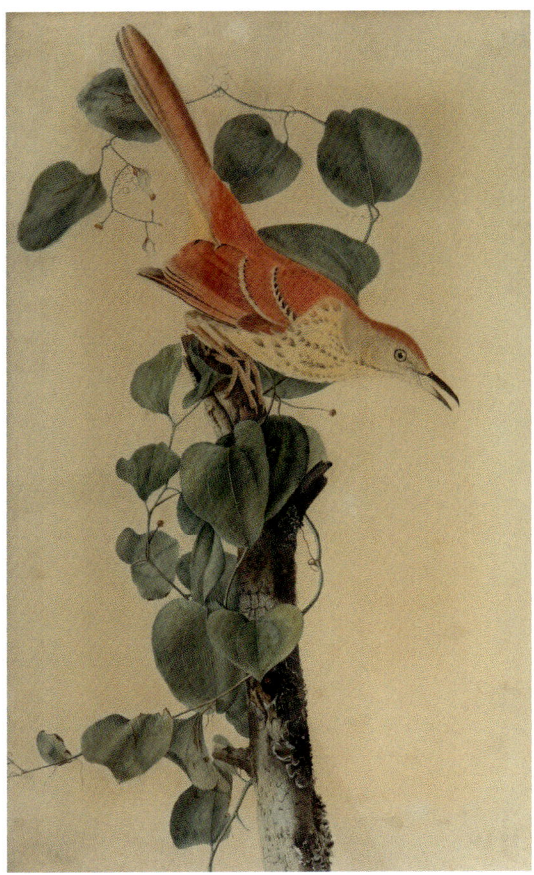

Audubon's depiction of multiple birds with cross-species interactions was innovative. However, Audubon's representation of birds in their natural habitats were influenced by the work of older naturalists like Mark Catesby. In this early watercolor, Audubon borrowed both the arrangement of the bird and the smilax vine from Catesby.

CAT 70 (ABOVE)
RED THRUSH OR BROWN THRASHER
John James Audubon

1815
Watercolor, graphite, and pastel on paper
John James Audubon State Park, Kentucky Department of Parks.

CAT 72 (RIGHT)
LETTER TO LUCY B. AUDUBON
John James Audubon

Liverpool, UK, November 25, 1827
Manuscript
APS.

Audubon found success in London by presenting himself not as an elite scientist, but as a rugged hunter from the American wilderness. Here he tells his wife Lucy that he longs for his wild 'universe' of America. He also asks for rattlesnake skins as presents for British supporters.

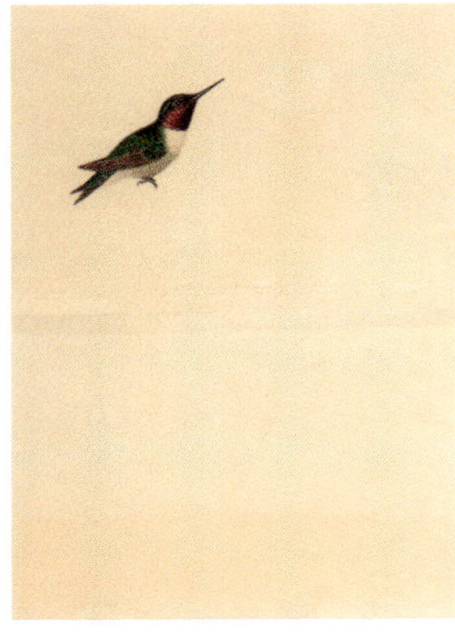

CAT 71
RUBY-THROATED HUMMINGBIRD
John James Audubon

c. 1821
Watercolor, pastel, ink, and graphite on paper
John James Audubon State Park, Kentucky Department of Parks.

Audubon produced accurate studies by impaling recently killed birds on a wire mounting system. He then manipulated the bodies into positions based on his field observations. Studies of individual birds like this one were later incorporated into finished compositions.

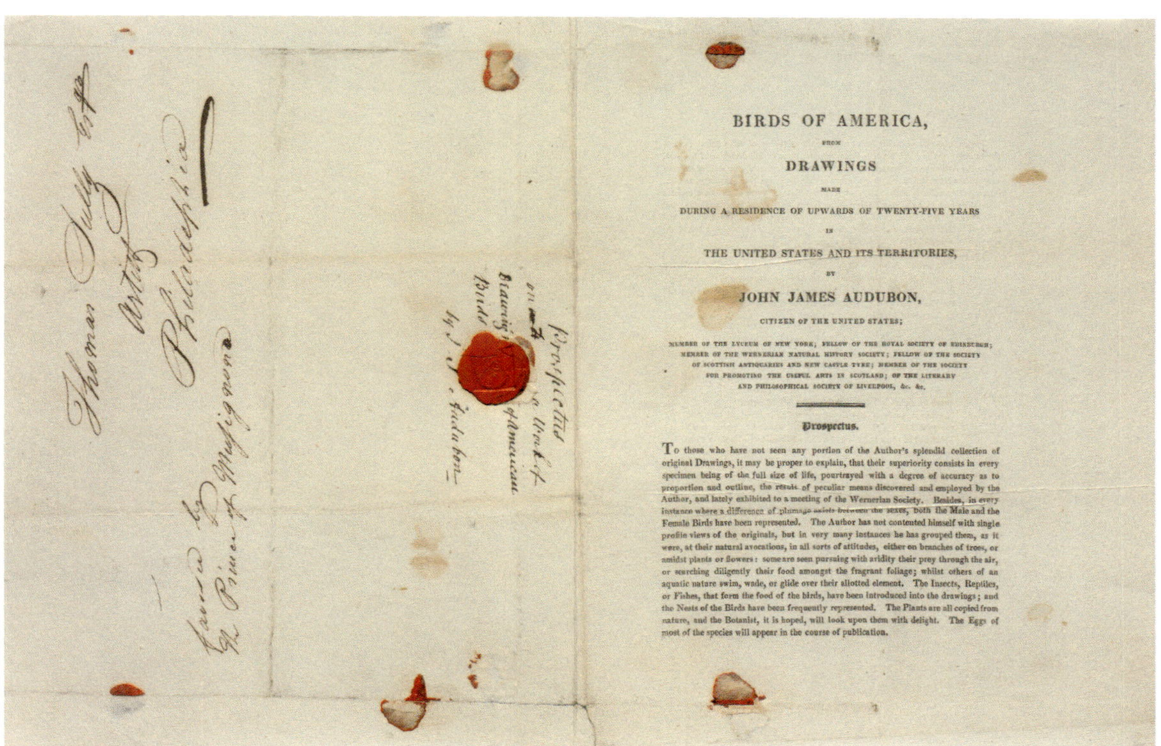

BIRDS OF AMERICA,

FROM

DRAWINGS

MADE

DURING A RESIDENCE OF UPWARDS OF TWENTY-FIVE YEARS

IN

THE UNITED STATES AND ITS TERRITORIES,

BY

JOHN JAMES AUDUBON,

CITIZEN OF THE UNITED STATES;

MEMBER OF THE LYCEUM OF NEW YORK; FELLOW OF THE ROYAL SOCIETY OF EDINBURGH; MEMBER OF THE WERNERIAN NATURAL HISTORY SOCIETY; FELLOW OF THE SOCIETY OF SCOTTISH ANTIQUARIES AND NEW CASTLE TYNE; MEMBER OF THE SOCIETY FOR PROMOTING THE USEFUL ARTS IN SCOTLAND; OF THE LITERARY AND PHILOSOPHICAL SOCIETY OF LIVERPOOL, &c. &c.

Prospectus.

To those who have not seen any portion of the Author's splendid collection of original Drawings, it may be proper to explain, that their superiority consists in every specimen being of the full size of life, pourtrayed with a degree of accuracy as to proportion and outline, the result of peculiar means discovered and employed by the Author, and lately exhibited to a meeting of the Wernerian Society. Besides, in every instance where a difference of plumage exists between the sexes, both the Male and the Female Birds have been represented. The Author has not contented himself with single profile views of the originals, but in very many instances he has grouped them, as it were, at their natural avocations, in all sorts of attitudes, either on branches of trees, or amidst plants or flowers: some are seen pursuing with avidity their prey through the air, or searching diligently their food amongst the fragrant foliage; whilst others of an aquatic nature swim, wade, or glide over their allotted element. The Insects, Reptiles, or Fishes, that form the food of the birds, have been introduced into the drawings; and the Nests of the Birds have been frequently represented. The Plants are all copied from nature, and the Botanist, it is hoped, will look upon them with delight. The Eggs of most of the species will appear in the course of publication.

Audubon's book proposal for *The Birds of America* makes bold claims for its originality. In particular, he highlights the fact that he illustrates birds at actual size—the reason for the volume's mammoth size— and contrasts his depiction of birds in their natural environment with previous publications.

CAT 73 (ABOVE)
THE BIRDS OF AMERICA . . . PROSPECTUS
John James Audubon

Edinburgh, 1827
Broadside
APS.

CAT 74 (RIGHT)
LETTER TO CHARLES WATERTON
George Ord

Philadelphia, April 23, 1832
Manuscript
APS.

Audubon's bid to create an improved American ornithology made him enemies. These included the naturalist George Ord, close friend of Alexander Wilson. Wilson was the author of the first definitive American ornithology. In this letter, Ord describes Audubon's volumes as 'vile.' He also questions their accuracy, correctly identifying one of Audubon's 'new' species as a fake.

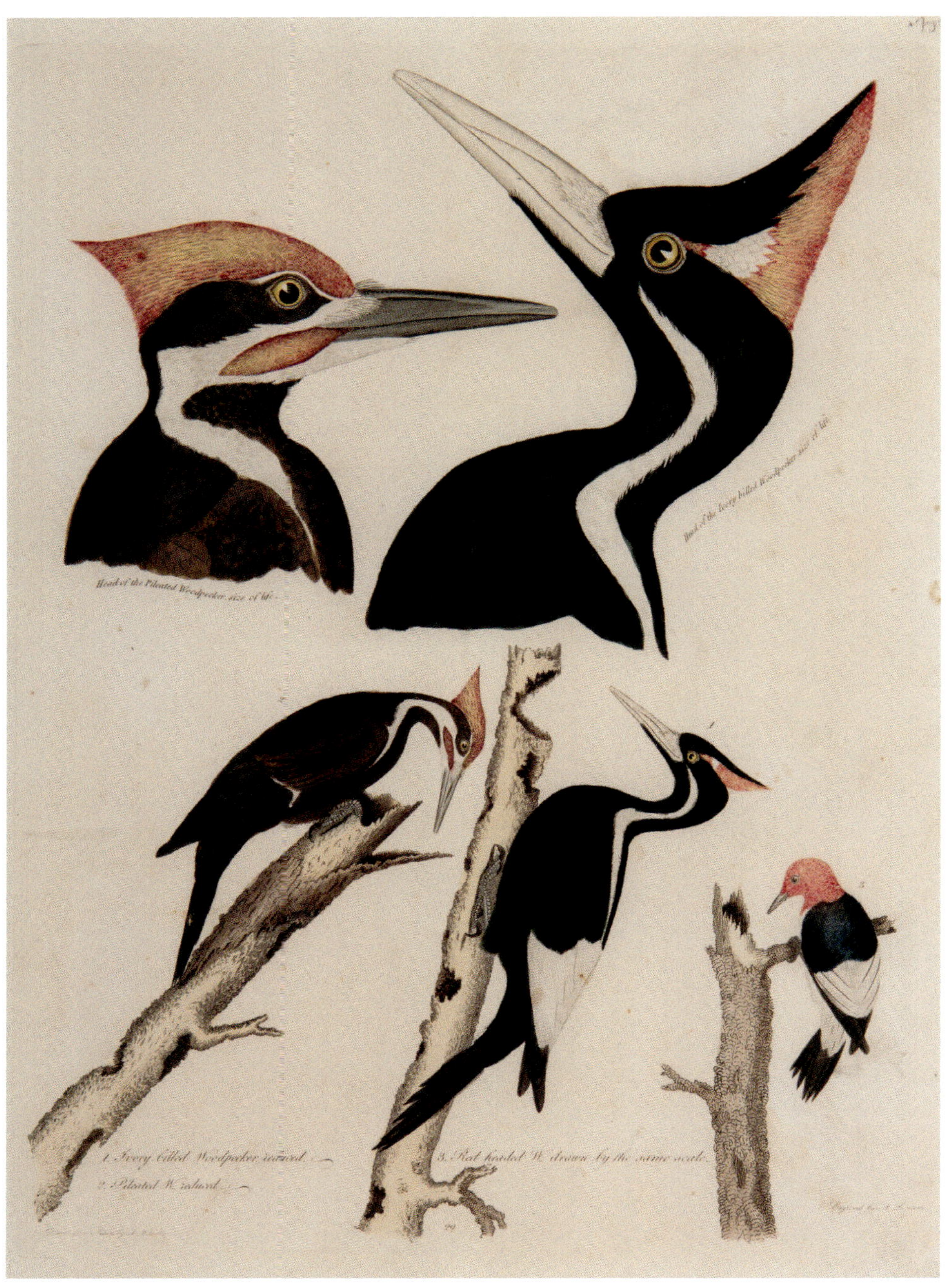

Audubon exaggerated the differences between his work and the work of previous naturalists, including Alexander Wilson. Wilson had already pioneered ornithological descriptions that included ample information about bird behavior. Wilson also corrected the misconception that woodpeckers destroy trees, when instead they are eating potentially harmful grubs.

CAT 75
IVORY-BILLED
WOODPECKER
Alexander Lawson, after
Alexander Wilson

Philadelphia, 1811
Hand-colored engraving
In *American Ornithology* . . .
vol. 4 by Alexander Wilson
Private Collection.

NATURALISTS COLLECTED THE HUMAN REMAINS of Native Americans and other socially marginalized groups on the very same expeditions that they collected American flora and fauna. Titian Ramsay Peale looted human remains from two Indigenous burial grounds during the Long Expedition. John James Audubon stole the remains of fifteen Native American, Mexican, and Hispano-Indian individuals. There are no documented instances in which William Bartram collected human remains, however, he looted Native American grave goods in Florida.

SCIENTISTS WHO WANTED TO SUPPORT FALSE CLAIMS of European biological superiority fueled the trade in human remains. In Philadelphia, the physician Samuel G. Morton (APS, 1828) assembled a vast collection of human skulls to support his racist theories. Both Peale and Audubon collected human remains for Morton.

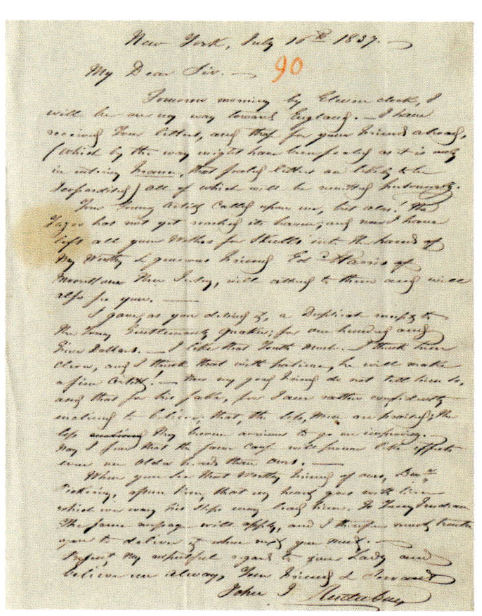

The works of Peale, Audubon, Bartram, and countless others reveal **COMPLICATED HISTORIES.** Philadelphia institutions, including the American Philosophical Society, endorsed the scientific pursuits of these individuals. While practices have changed over time, scientists and institutions still grapple with this question today—**HOW HAS SCIENCE SUPPORTED INJUSTICE?**

Audubon collected the remains of about 15 Native American, Hispano-Indian, and Mexican people. He sent the remains of 10 individuals to Samuel G. Morton. Here Audubon discusses the remains of a Yazoo individual with Morton. The Yazoo were a small tribe in present-day Mississippi that joined the Chickasaw and Choctaw around the 1730s.

CAT 76
LETTER TO SAMUEL G. MORTON
John James Audubon
New York, July 16, 1837
Manuscript
APS.

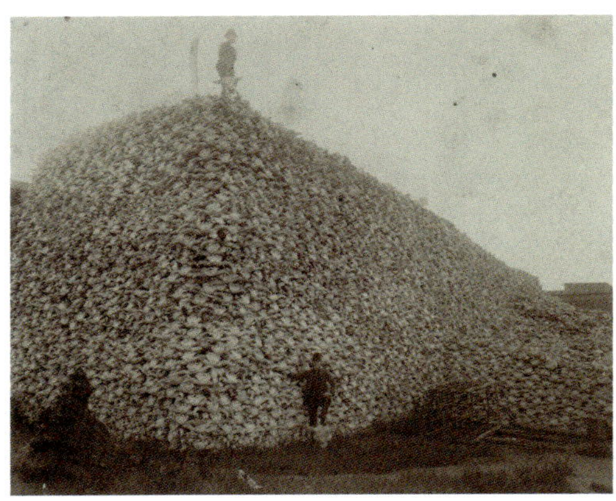

FIG 20 MAN STANDS ON TOP OF ENORMOUS PILE OF BUFFALO (BISON) SKULLS
Unknown Photographer, c. 1892, Photographic print mounted on mat board, Courtesy of the Burton Historical Collection, Detroit Public Library.

Market-driven hunting of the American bison nearly pushed the species to extinction in the late 19th century.

LANDSCAPES OF CHANGE

The Quadrupeds of North America (1845–1848) was John James Audubon's last major project. His health was failing and much of the work would be completed by his two sons and his collaborator, John Bachman. Work began at a time when human-caused environmental change had accelerated in North America. A large number of images in *Quadrupeds* depict environmental destruction, as animals are haunted by fields filled with tree stumps and fences. Audubon rarely, however, protested environmental destruction in public. He was also quick to justify his own hunting—up to one hundred birds per day in the field—as necessary death in the pursuit of science.

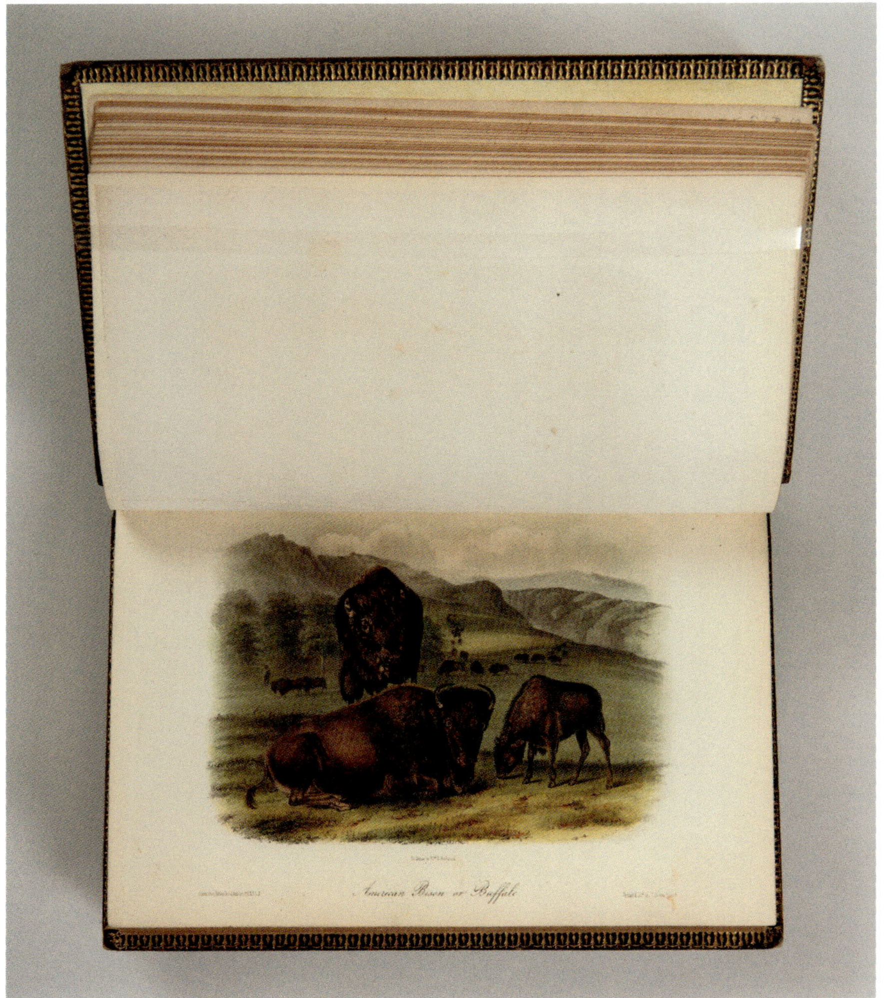

The essay that accompanies this image describes the bison as one of the last remaining links to America's extinct megafauna, in particular the mastodon, and suggests that it might be next. These warnings about extinction were not present in Audubon's draft and were probably added by John Bachman.

CAT 77
THE QUADRUPEDS OF NORTH AMERICA, VOL. 2
John James Audubon and John Bachman

New York, 1854 (2nd edition, Octavo version)
Hand-colored lithograph
APS.

Some writers, as Morse, &c. assert, that the horns of the Elk are shed every year, in the month of February, and that by august the new horns are nearly arrived at their full growth. This, however, is a vulgar opinion, by no means founded in truth.

Within the memory of many persons, upon whose authority I have full dependance, the droves of elks which used to frequent the salines near the river Susquehanna, &c. were so great that, for five or six miles, leading to these salines, their paths, or roads, were as large as many of the great public roads of our country. Eighty elks have sometimes been seen in one herd, upon their march to the salines. But the astonishing progress of population in our infant countries has greatly changed the scene. Nothing but the establishment of laws, for the purpose, will prevent many of our animals [in a few years] from being almost entirely exterpated, through the whole of that extensive tract of territory within the limits of the United-States. The bison, the elk, and the common deer were, a few years since, as common animals on the Atlantic-side of the United-States as they are now in the almost unexplored and even unvisited coun—

As early as the 1790s, Benjamin Smith Barton expressed fears about rapidly diminishing North American animal populations due to the country's growth, and suggested the need for legislation. Here he argued that bison, elk, deer, and beaver were particularly at risk.

CAT 78
ON THE ANIMALS OF NORTH AMERICA
Benjamin Smith Barton

c. 1793
Manuscript
APS.

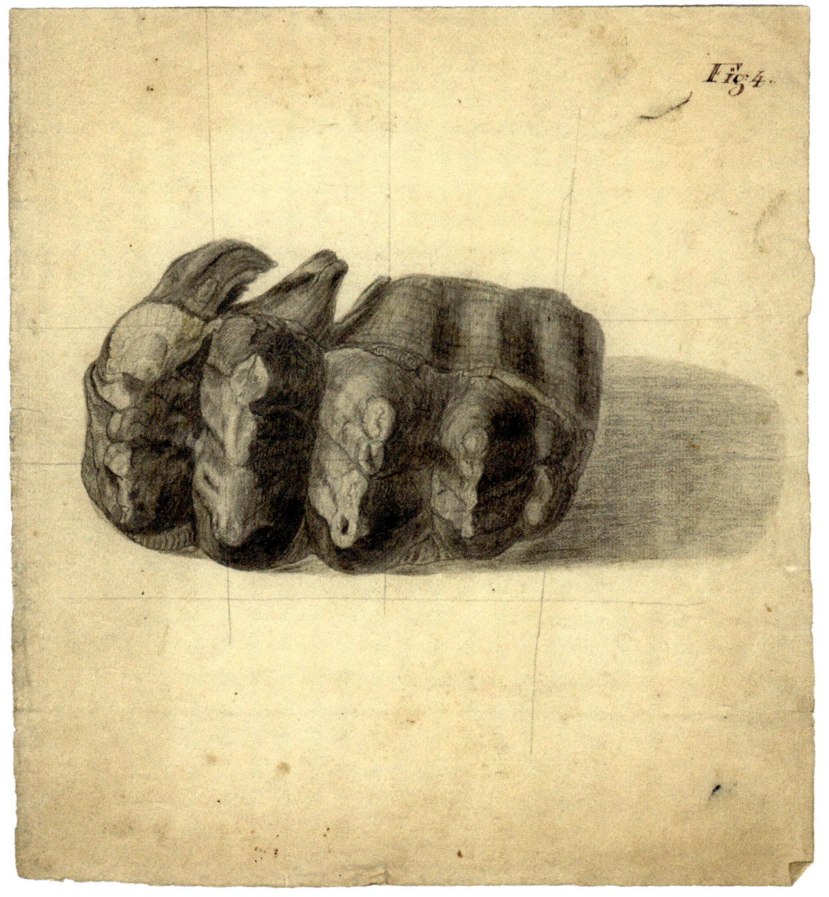

Many 18th-century scientists believed it was not possible for an entire species to die out. Scientific advancements around 1800 made it possible to prove the reality of extinction using comparisons between fossilized bones and bones of living species. A French comparative anatomist called Georges Cuvier used samples of teeth and other bones to distinguish the extinct American mastodon both from another extinct species—the mammoth—and from modern elephants, publishing the results in 1806. These discoveries set the stage for understanding humanity's role in extinction.

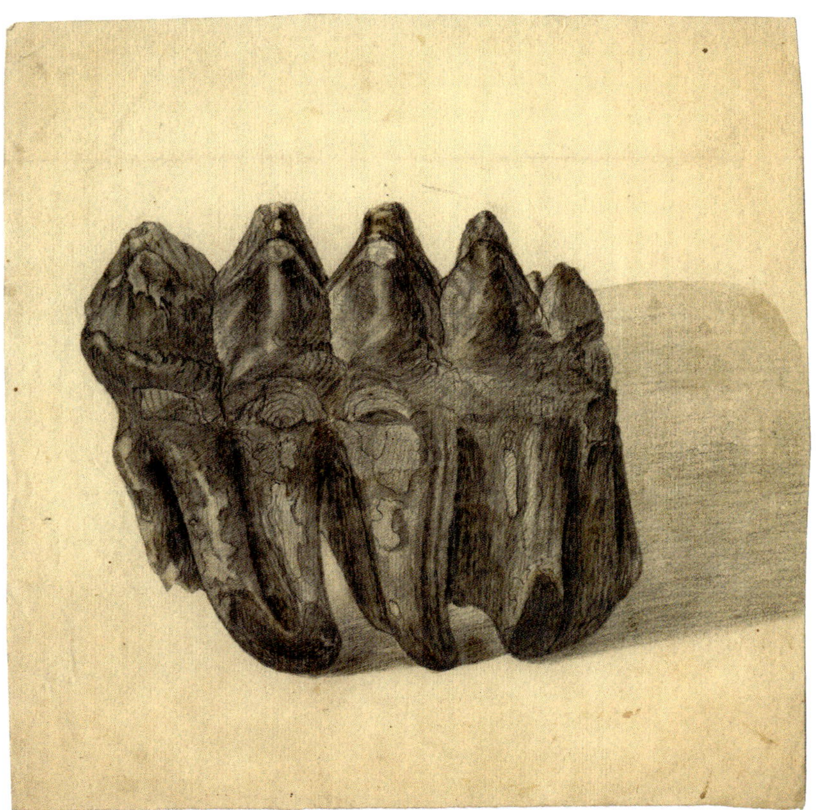

CAT 79
MASTODON
TOOTH [MOLAR],
CROWN VIEW
Benjamin Smith Barton
N.D.
Graphite on paper
APS.

CAT 80
MASTODON
TOOTH [MOLAR],
LATERAL VIEW
Benjamin Smith Barton
N.D.
Graphite on paper
APS.

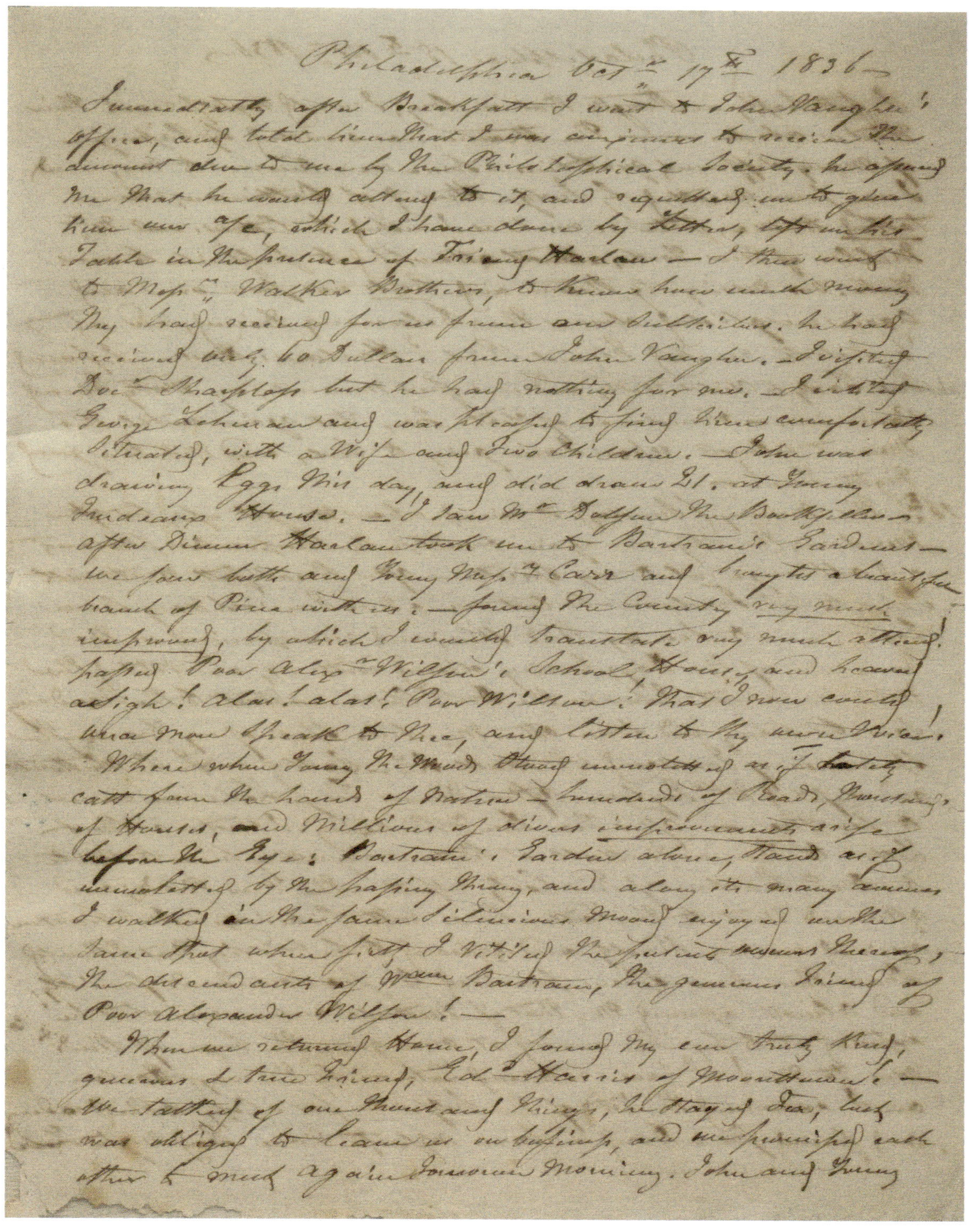

Audubon describes a visit to Bartram's Garden in 1836. He mourns the death of Alexander Wilson and also offers an ambivalent description of the changes to an area that had once seemed, "unmolested as if lately cast from the hand of Nature."

CAT 81
JOURNAL
John James Audubon
Philadelphia, October 17, 1836
Manuscript
APS.

AUDUBON'S UNACKNOWLEDGED COLLABORATORS

The Birds of America and *The Quadrupeds of North America* were collaborative projects. They required observing animals across North America, producing thousands of images, and writing volumes of text. Among Audubon's collaborators were many women, Native Americans, and African laborers who made Audubon's work possible. Maria Martin, the sister-in-law and later wife of Audubon's collaborator John Bachman, produced botanical images for many plates in *Birds*. A man named Thomas, whom Bachman enslaved, taxidermied specimens during the years Audubon collaborated with Bachman on *Birds*. Finally, Natoyist-Siksina' (Medicine Snake Woman) was a Kainai woman who helped facilitate the multiracial knowledge networks on which Audubon relied.

Martin painted the botanical, a Franklin tree, in the original study for *Birds*, to which Audubon added the Bachman's Warblers. Martin was responsible for botanicals and insects in approximately 30 compositions for *Birds*.

CAT 82
BACHMAN'S WARBLER,
STUDY FOR HAVELL PL. 185
Oppenheimer Editions, after
John James Audubon and
Maria Martin

2006 (Original 1833)
Fine Art Print (Original
watercolor, graphite, gouache,
and black ink on paper)
APS. (Watercolor in the
Collection of the New-York
Historical Society Museum &
Library)

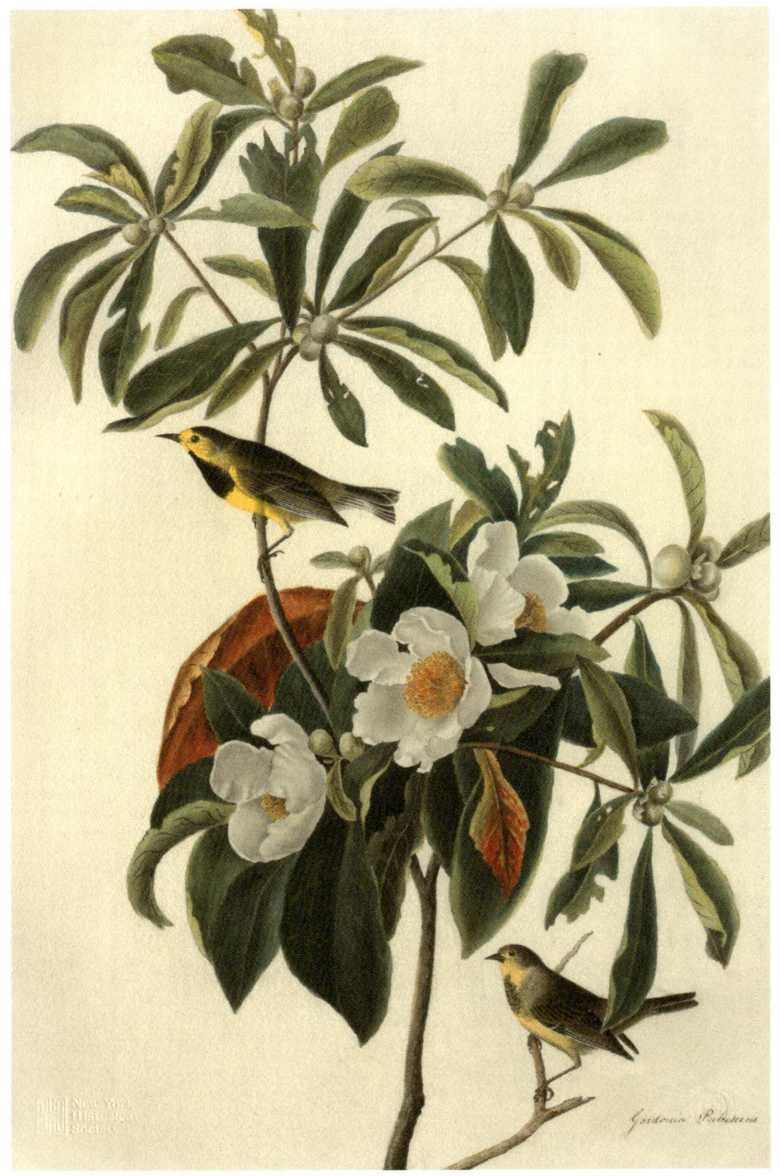

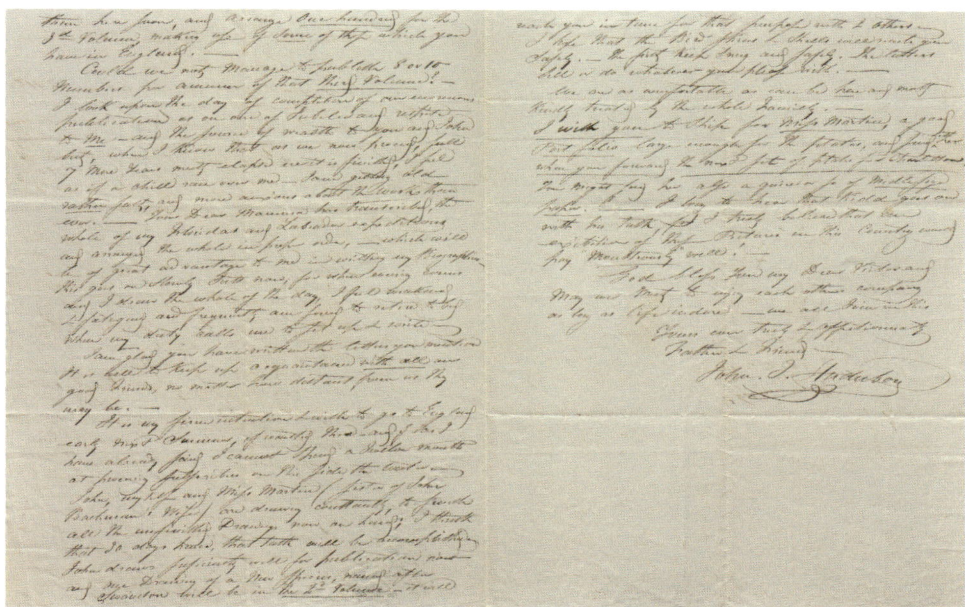

Audubon met naturalist John Bachman and Maria Martin in the fall of 1831 and quickly discovered that Martin had a talent for painting plants. In this letter, Audubon describes a visit to the Bachmans. He, his son John Woodhouse, and Martin were all hard at work finishing studies for *The Birds of America.*

Audubon named this warbler for the naturalist John Bachman, brother-in-law and later husband of Maria Martin. Bachman sent Audubon specimens and descriptions of southern birds, including this warbler, which Audubon never saw in the wild. Both Bachman and Martin were enslavers, and Bachman was a vocal anti-abolitionist.

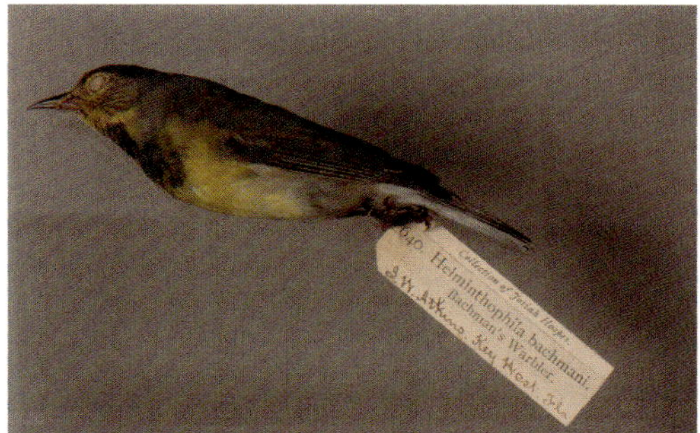

CAT 84
BACHMAN'S WARBLER
Collected and prepared by
S.W. Atkins

August 11, 1890
Ornithological specimen
The Academy of Natural
Sciences of Drexel University.

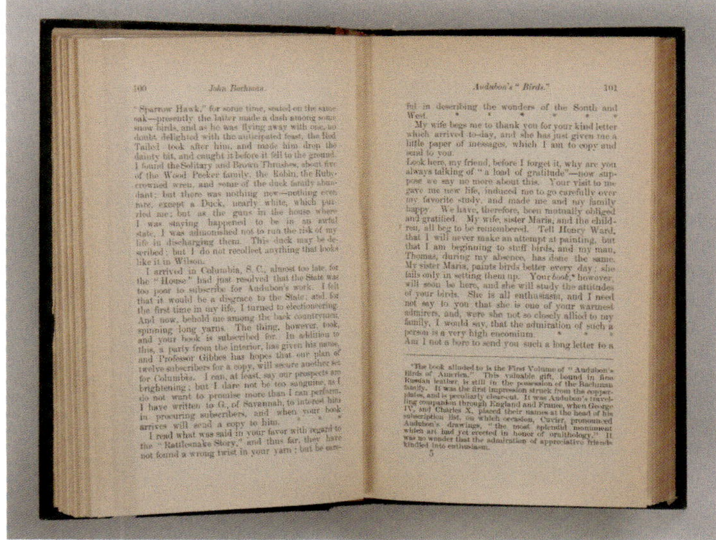

CAT 85
JOHN BACHMAN: THE PASTOR OF SAINT JOHN'S LUTHERAN CHURCH, CHARLESTON
Catherine L. Bachman

Charleston, 1888
Bound volume
APS. Presented by Samuel Henshaw.

Bachman tells Audubon that his first visit inspired the whole Bachman household to take up ornithology. He explicitly notes that Thomas, a man enslaved by Bachman, had begun to taxidermy birds. Thomas's specific contributions cannot be identified, but Audubon visited the Bachmans multiple times while preparing *Birds* and Bachman sent him many taxidermied bird specimens.

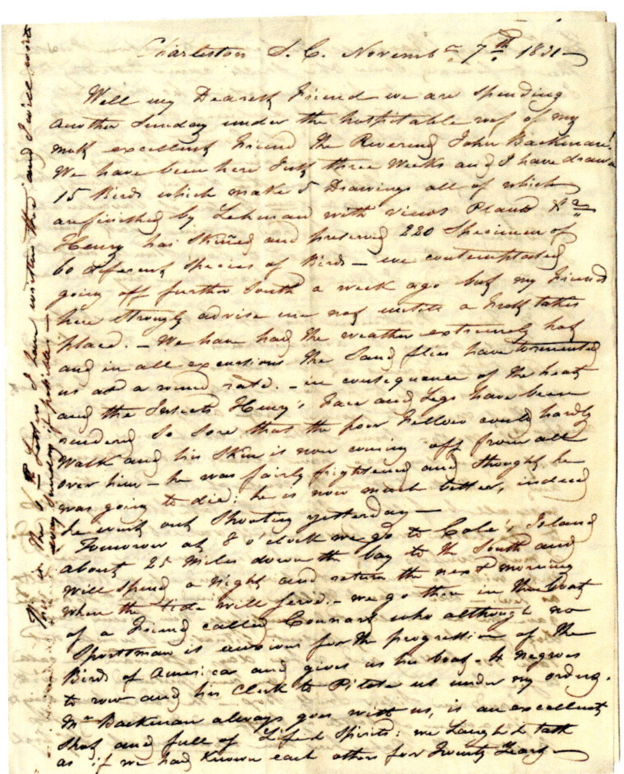

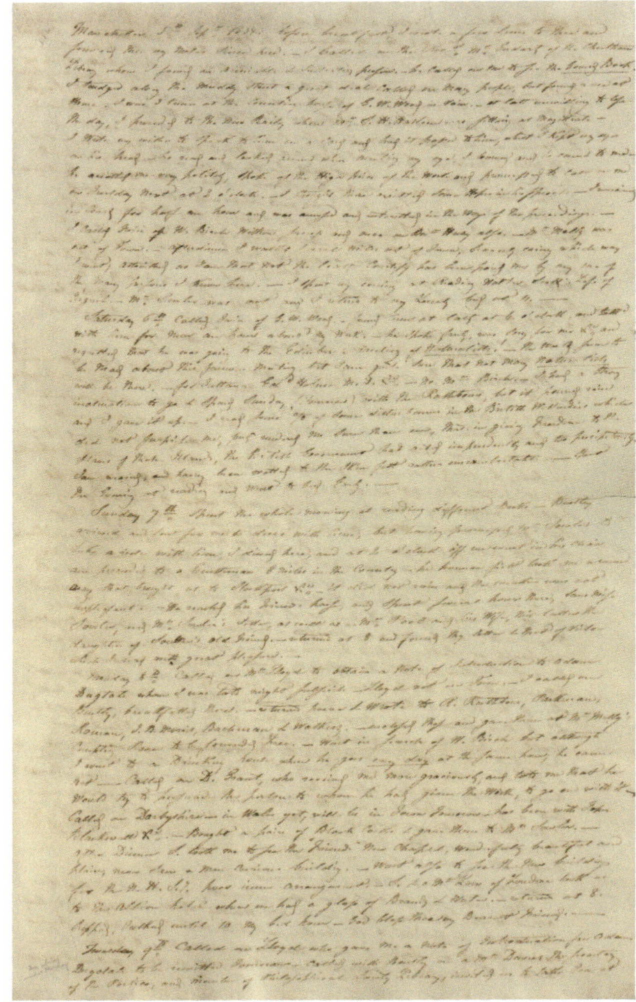

Audubon relied on the contributions of countless enslaved individuals while working on *The Birds of America*. This letter describes an excursion to an island near Charleston where he was accompanied by four enslaved people. It is one of many such hints in Audubon's correspondence that do not appear in his published writings.

CAT 86
LETTER TO LUCY
BAKEWELL AUDUBON
John James Audubon

Charleston, November 7, 1831
Manuscript
APS.

John James Audubon's position on slavery is well-documented. He enslaved people during the 1810s while living in Henderson, Kentucky, and again in the 1820s. Moreover in September 1834, he wrote disapprovingly to his wife Lucy about Parliament's recent decision to emancipate enslaved people in the British West Indies.

CAT 87
LETTER TO LUCY
BAKEWELL AUDUBON
John James Audubon

Manchester, UK,
September 5–11, 1834
Manuscript
APS.

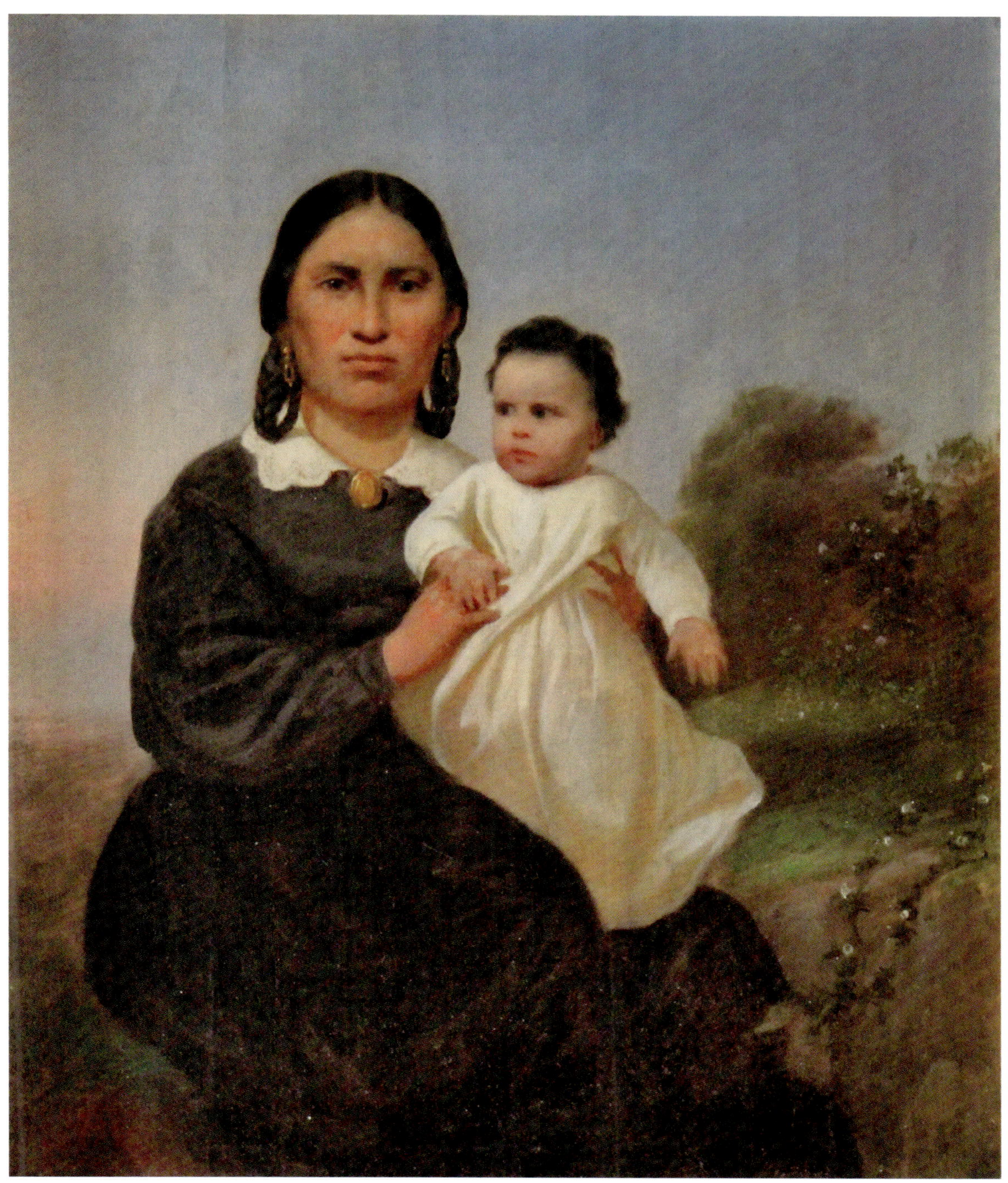

Audubon stayed at Fort Union (North Dakota/ Montana border) in summer 1843 while on an expedition for *The Quadrupeds of North America*. There, he met Natoyist-Siksina', a Kainai (Blackfoot Confederacy) woman married to Alexander Culbertson, the fur trader who ran the fort. Natoyist-Siksina's diplomacy was essential to maintaining good relations with the Native American hunters and guides on whom Audubon relied.

CAT 88
**NATOYIST-SIKSINA'
(MEDICINE SNAKE
WOMAN; MRS. ALEXANDER
CULBERTSON)**
John James Audubon

c. 1843
Oil on canvas
John James Audubon State
Park, Kentucky Department
of Parks.

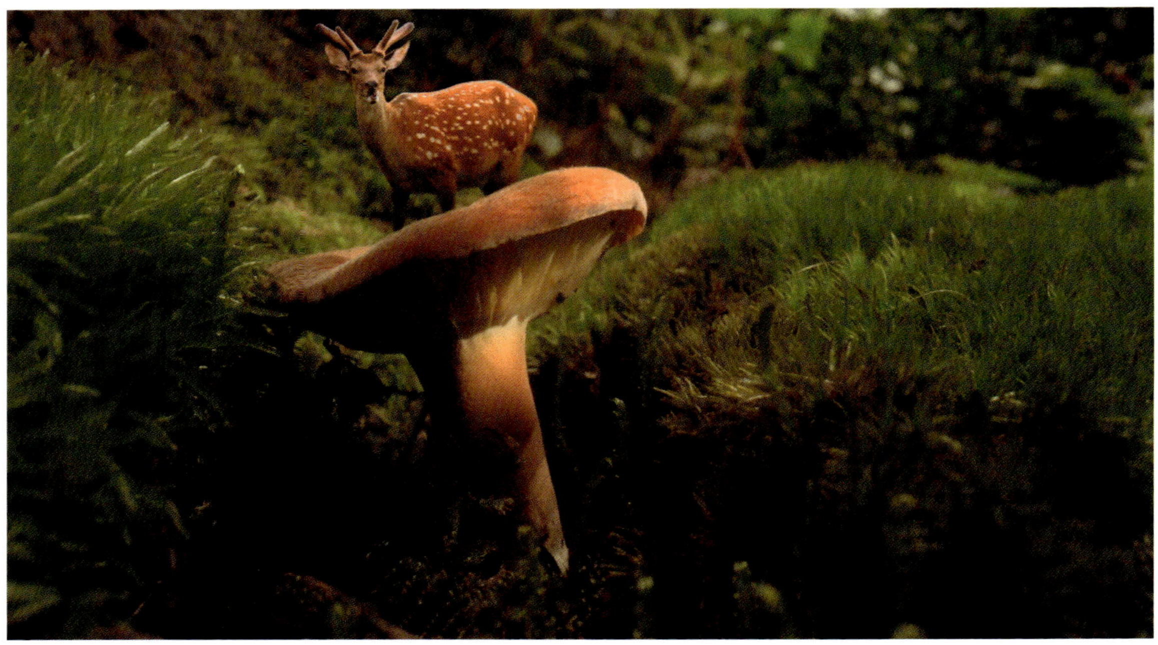

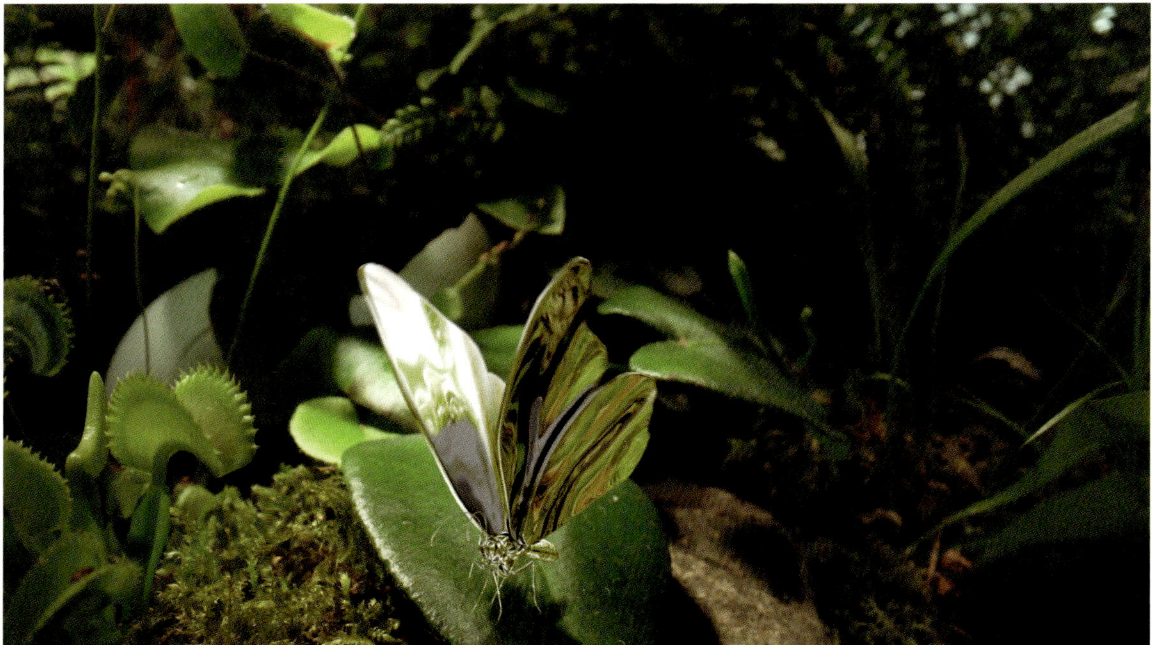

Set in a microcosmic forest, *Field Companion* is based loosely on the pine barrens that dot Southern New Jersey near the home of Hironaka & Suib. The film considers forest ecosystems in terms of symbiotic and collaborative relationships that sustain coexistence and community. In *Field Companion*, the pine barrens forest is condensed and transplanted to a terrarium in the artists' studio. The terrarium's diverse native plants and living dwellers are accompanied by digitally rendered part-animal, part-human creatures. Through these dwellers' conversations and interactions, they look forward, investigating progressive methods of sustainability for now and into the future.

CAT 89
FIELD COMPANION
Nadia Hironaka and
Matthew Suib

2021
4k video with sound, RT: 20:40
Co-commissioned in 2021 by
Rowan University Art Gallery
and Locust Projects, Miami
Courtesy the artists and
Locks Gallery, Philadelphia.

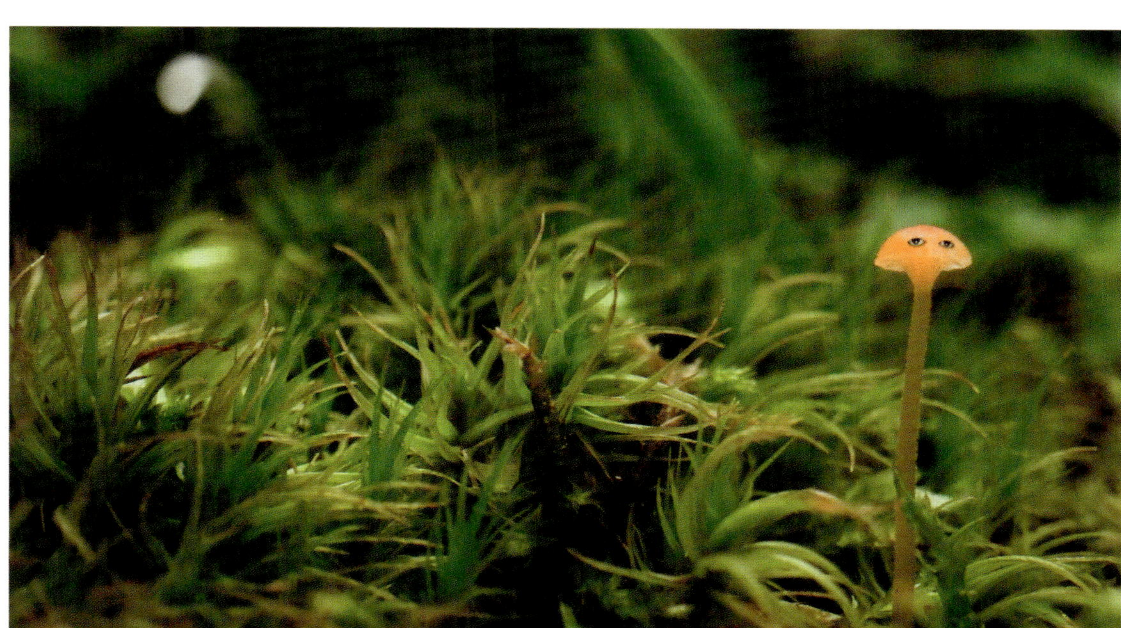

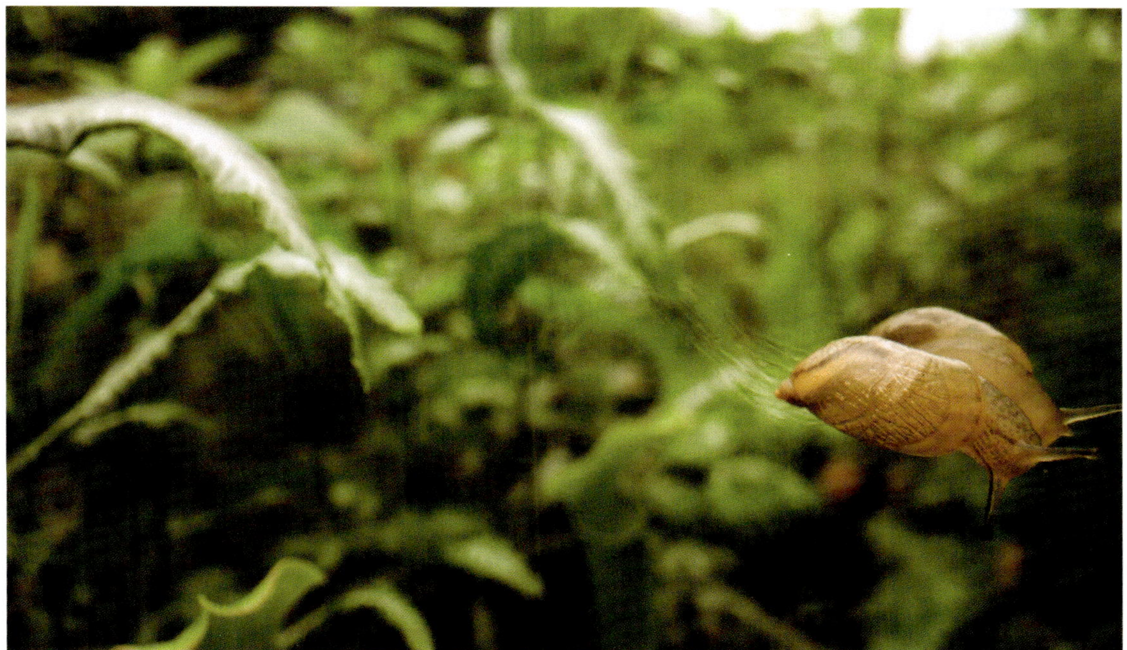

Artists today continue to explore biodiversity and ecology. In *Field Companion*, Nadia Hironaka and Matthew Suib use film and digital tools to represent the natural world, rather than pencil, watercolor, and paper. Yet, many of the questions they ask are ones that already emerged in the work of Bartram, Peale, and Audubon—how can art capture nature's complexity? How are humans tied to the rest of the natural world? What do we learn when we look at the world from a non-human perspective?

Athens, Elizabeth. "Figuring a World: *William Bartram's* Natural History." PhD diss., Yale University, 2015.

Audubon, John James. *Ornithological Biography.* 5 vols. Edinburgh: Adam Black, 1831–1839.

——————. *John James Audubon: Writings and Drawings.* Edited by Christoph Irmscher. New York: Literary Classics of the United States, 1999.

Bartram, William. *The* Travels *of William Bartram: Naturalist's Edition.* Edited by Francis Harper. New Haven: Yale University Press, 1958.

Blakney, Sharece. *Stories We Know: Recording the Black History of Bartram's Garden and Southwest Philadelphia.* Edited by Aislinn Pentecost-Farren. Philadelphia: John Bartram Association/ Mural Arts Philadelphia, 2017.

Blaugrund, Annette and Theodore E. Stebbins, Jr., eds. *John James Audubon: The Watercolors for* The Birds of America. New York: The New-York Historical Society, 1993.

Blum, Ann Shelby. *Picturing Nature: American Nineteenth-century Zoological Illustration.* Princeton: Princeton University Press, 1993.

Boldt, Janine Yorimoto. *Dr. Franklin: Citizen Scientist.* Philadelphia, PA: American Philosophical Society Press, 2020.

Braund Kathryn H., ed. *The Attention of a Traveller: Essays on William Bartram's* Travels *and Legacy.* Tuscaloosa: The University of Alabama Press, 2022.

Brigham, David R. *Public Culture in the Early Republic: Peale's Museum and Its Audience.* Washington, DC: Smithsonian Institution Press, 1995.

Drayton, Richard. *Nature's Government: Science, Imperial Britain, and the 'Improvement' of the World.* New Haven: Yale University Press, 2000.

Ewan, Joseph. *William Bartram: Botanical and Zoological Drawings, 1756–1788.* Philadelphia: The American Philosophical Society, 1968.

Farber, Paul L. *Discovering Birds: The Emergence of Ornithology as a Scientific Discipline, 1760–1850.* Baltimore and London: Johns Hopkins University Press, 1997.

Fernandez-Sacco, Ellen. "Framing 'The Indian': The Visual Culture of Conquest in the Museums of Pierre Eugene Du Simitiere and Charles Willson Peale, 1779–96." *Social Identities* 8 (2002): 571–618.

Fry, Joel T. "Slavery and Freedom at Bartram's Garden." Paper presented at the Investigating Mid-Atlantic Plantations: Slavery, Economies, and Space Conference, Philadelphia, Pennsylvania, October 2019. https://www.bartramsgarden. org/wp-content/uploads/J.Fry-slavery-freedom-at-Bartrams-Garden-MCEAS-2019-4.pdf.

Fry, Joel T. and Amy Meyers. "William Bartram (1739–1823), *Catalogue of American Trees, Shrubs and Herbaceous Plants, most of which are now growing, and produce ripe Seed in John Bartram's Garden, near Philadelphia.*" In *Two Hundred Years: The Historical Society of Pennsylvania, 1824–2024.* Edited by David R. Brigham, 92–94. Philadelphia: The Historical Society of Pennsylvania, 2023.

Gaudio, Michael. "Swallowing the Evidence: William Bartram and the Limits of Enlightenment." *Winterthur Portfolio* 36 (Spring 2001): 1–17.

Hall, Ryan. *Beneath the Backbone of the World: Blackfoot People and the North American Borderlands, 1720–1877.* Chapel Hill: The University of North Carolina Press, 2020.

Halley, Matthew R. "Audubon's Bird of Washington: Unravelling the Fraud that launched *The Birds of America.*" *Bulletin of the British Ornithologists' Club* 140 (2020): 110–141.

Hallock, Thomas. "'On the Borders of a New World': Ecology, Frontier Plots, and Imperial Elegy in William Bartram's 'Travels'." *South Atlantic Review* 66 (Autumn 2001): 109–133.

Hallock, Thomas and Nancy E. Hoffman, eds. *William Bartram, the Search for Nature's Design: selected art, letters, and unpublished writings.* Athens, GA: The University of Georgia Press, 2010.

Haltman, Kenneth. *Looking Close and Seeing Far: Samuel Seymour, Titian Ramsay Peale, and the Art of the Long Expedition, 1818–1823.* University Park, PA: Pennsylvania State University Press, 2008.

Iannini, Christopher. *Fatal Revolutions: Natural History, West Indian Slavery, And the Routes of American Literature.* Chapel Hill: Published for the Omohundro Institute of Early American History and Culture, Williamsburg, Virginia, by the University of North Carolina Press, 2012.

Irmscher, Christoph. *The Poetics of Natural History.* With Rosamond Purcell. 2nd edition. New Brunswick, NJ: Rutgers University Press, 2019.

James, Edwin. *Account of an Expedition from Pittsburgh to the Rocky Mountains, Performed in the Years 1819 and '20, By Order of The Hon. J.C. Calhoun, Sec'y of War, Under the Command of Major Stephen H. Long. From the Notes of Major Long, Mr. T. Say, and Other Gentlemen of the Exploring Party.* 2 vols. Philadelphia: H.C. Carey and I. Lea, 1822.

Jones, E. Bennett. "'The Indians Say': Settler Colonialism and the Scientific Study of North America, 1722 to 1848." PhD diss., Northwestern University, 2021.

Judd, Richard W. *The Untilled Garden: Natural History and the Spirit of Conservation in America, 1740–1840.* New York: Cambridge University Press, 2009.

Lanham, J. Drew. "What Do We Do About John James Audubon?" *Audubon Magazine* (Spring 2021), https://www.audubon.org/magazine/spring-2021/what-do-we-do-about-john-james-audubon.

Lewis, Andrew J. "A Democracy of Facts, an Empire of Reason: Swallow Submersion and Natural History in the Early American Republic." *William and Mary Quarterly* 62 (October 2005): 663–696.

Lindsay, Debra J. *Maria Martin's World: Art & Science, Faith & Family in Audubon's America.* Tuscaloosa: The University of Alabama Press, 2018.

Looby, Christopher. "The Constitution of Nature: Taxonomy as Politics in Jefferson, Peale, and Bartram." *Early American Literature* 22 (1987): 252–273.

Magee, Judith. *The Art and Science of William Bartram.* University Park, PA: Pennsylvania State University Press, 2007.

Messer, Peter C. "The Nature of William Bartram's *Travels.*" In *Atlantic Environments and the American South.* Edited by Thomas Blake Earle and D. Andrew Johnson, 195–214. Athens, GA: University of Georgia Press, 2020.

Meyers, Amy R.W. "Sketches from the Wilderness: Changing Conceptions of Nature in American Natural History Illustration: 1680–1880." PhD diss., Yale University, 1985.

Meyers, Amy R.W. with Lisa L. Ford, eds. *Knowing Nature: Art and Science in Philadelphia, 1740–1840.* New Haven and London: Yale University Press, 2011.

Murphy, Robert Cushman. "The Sketches of Titian Ramsay Peale (1799–1885)." *Proceedings of the American Philosophical Society* 101 (Dec. 19, 1957): 523–531.

Nobles, Gregory. *John James Audubon: The Nature of the American Woodsman.* Philadelphia: University of Pennsylvania Press, 2017.

Olson, Roberta J.M with Marjorie Shelley and Alexandra Mazzitelli. *Audubon's Aviary: The Original Watercolors for* The Birds of America. New York: New-York Historical Society, 2012.

Parrish, Susan Scott. *American Curiosity: Cultures of Natural History in the Colonial British Atlantic World.* Chapel Hill, NC: Published for the Omohundro Institute of Early American History and Culture, Williamsburg, Virginia, by the University of North Carolina Press, 2006.

Parsons, Christopher M. *A Not-So-New World: Empire and Environment in French Colonial North America.* Philadelphia: University of Pennsylvania Press, 2018.

Partridge, Linda Dugan. "By the Book: Audubon and the Tradition of Ornithological Illustration." In *Art and Science in America: Issues of Representation.* Edited by Amy R.W. Meyers, 97–130. San Marino, CA: Henry E. Huntington Library and Art Gallery, 1998.

Patterson, Daniel. "Audubon's Conservation Ethic Reconsidered." In *The Missouri River Journals of John James Audubon,* 211–304. Lincoln and London: University of Nebraska Press, 2016.

Poesch, Jessie. *Titian Ramsay Peale, 1799–1885, And His Journals of the Wilkes Expedition.* Philadelphia: American Philosophical Society, 1961.

Porter, Charlotte M. "The Lifework of Titian Ramsay Peale." *Proceedings of the American Philosophical Society* 129 (1985): 300–312.

Pratt, Mary Louise. *Imperial Eyes: Travel Writing and Transculturation.* 2nd ed. London and New York: Routledge, 2008.

Prince, Sue Ann, ed. *Stuffing Birds, Pressing Plants, Shaping Knowledge: Natural History in North America, 1730–1860.* Philadelphia, PA: American Philosophical Society, 2003.

Roberts, Jennifer L. *Transporting Visions: The Movement of Images in Early America.* Berkeley, Los Angeles, London: University of California Press, 2014.

Schiebinger, Londa L. *Plants and Empire: Colonial Bioprospecting in the Atlantic World.* Cambridge, MA: Harvard University Press, 2004.

Sellers, Charles Coleman. *Mr. Peale's Museum: Charles Willson Peale and the First Popular Museum of Natural Science and Art.* New York: W.W. Norton & Company, Inc., 1980.

Stearns, Raymond Phineas. *Science in the British Colonies of America.* Chicago: University of Illinois Press, 1970.

Strang, Cameron B. *Frontiers of Science: Imperialism and Natural Knowledge in the Gulf South Borderlands, 1500–1850.* Chapel Hill: University of North Carolina Press, 2018.

Waselkov, Gregory and Kathryn E. Holland Braund. *William Bartram on the Southeastern Indians.* Lincoln, NE: University of Nebraska Press, 1995.

Weese, A.O., ed. "The Journal of Titian Ramsay Peale." *Missouri Historical Review* 41 (January 1947): 147–163, (April 1947): 266–284.

Welch, Margaret. *The Book of Nature: Natural History in the United States, 1825–1875.* Boston: Northeastern University Press, 1998.